A-Z
GENETICS

A-Z GENETICS

Prof. Nirmal Chandra Pradhan

CENTRUM PRESS
NEW DELHI-110002 (INDIA)

CENTRUM PRESS

H.O.: 4360/4, Ansari Road, Daryaganj,
New Delhi-110 002 (India)
Ph.: 23278000, 23261597

B.O.: No. 1015, Ist Main Road, BSK IIIrd Stage
IIIrd Phase, IIIrd Block,
Bangalore - 560 085 (India)
Tel.: 080-41723429
Visit us at: www.centrumpress.com

A-Z Genetics

First Edition, 2009

ISBN 978-93-80106-22-9

PRINTED IN INDIA

Printed at Salasar Imaging Systems, Delhi-110035 (India)

Contents

Preface

Genetics a discipline of biology is the science of heredity and variation in living organisms. The fact that living things inherit traits from their parents has been used since prehistoric times to improve crop plants and animals through selective breeding. However, the modern science of genetics, which seeks to understand the process of inheritance, only began with the work of Gregor Mendel in the mid-nineteenth century.

Genes correspond to regions within DNA, a molecule composed of a chain of four different types of nucleotides the sequence of these nucleotides is the genetic information organisms inherit. The sequence of nucleotides in a gene is translated by cells to produce a chain of amino acids, creating proteins the order of amino acids in a protein corresponds to the order of nucleotides in the gene.

The amino acids in a protein determine how it folds into a three-dimensional shape this structure is in turn responsible for the protein's function. Proteins carry out almost all the functions needed for cells to live. A change to the DNA in a gene can change a protein's amino acids, changing its shape and function. Although genetics plays a large role in the appearance and behavior of organisms, it is the combination of genetics with what an organism experiences that determines the ultimate outcome.

Author

Chapter 1

Fundamentals of Genetic

Behaviour genetics from where do diverse scientific disciplines emerge, develop, and finally coalesce? Psychology, which attempts the description, prediction, and ultimately the understanding of Behaviour, began in the universities of Europe. One tradition emerged from the 18th and 19th-century German philosophers as exemplified by Leibniz, and was given an experimental basis by Wundt. The other, emanating from England, was based on the principles of John Locke and the *philosophes* of the French enlightenment.

This Anglo French influence came to America in 1882 with Herbert Spencer, who was enamored of Charles Darwin's theory of natural selection. Although these traditions are diffuse they help provide an orientation for understanding the development of psychology and its attempts to provide a firm biological basis for the study of Behaviour. The Lockean prescription offered a more comfortable cultural fit to our society, and its view of mental process as the product of nothing but sensation furnished a comfortable ambience for the growth.

PRINCIPLES OF GENETICS

MENDELIAN GENETICS

Genetics formal beginnings are found in the work of Gregor Mendel whose elegant experiments with garden peas and mice provided the foundation for the science of genetics. In the garden of a monastery in Moravia (now Czechoslovakia), of which he eventually became abbot,

Mendel crossbred varieties of the garden pea that had maintained relatively constant differences in physical characteristics over generations. This hereditary constancy occurred despite the occasional, now fully scientifically explained, disappearance of a given trait during several generations and its subsequent reemergence.

Some of the factors Mendel analyzed were tall versus dwarf plants, red or white colored blossoms, seed colour and shape, pod forms, and the position of flowers on stems. He theorized that each plant contained two factors for a particular characteristic such as colour (one of each of these factors contributed by each parental plant) and that these units were transmitted to descendants in adherence to statistical predictions.

These two factors (now called genes) might be identical, in which case the organism was said to be homozygous for that gene, or they might differ and the organism would be heterozygous.

From his plant breeding experiments and observations, Mendel derived two laws that serve as the foundation, both historically and functionally, of the science of genetics. The first law concerns segregation. Genes are particulate units of inheritance that may exist in two or more variant forms (alleles) at the same position (locus) on a chromosome. Members of a chromosome pair separate when sex cells are produced. If an individual is heterozygous at the *A* locus, the gametes will be of two types—A or *a*.

The second law deals with independent assortment of alleles on different chromosomes. When the chromosomes separate to form eggs and sperm, these variant alleles (*Aa* and *Bb*, for example) are transmitted independently of one another to the next generation. When one allele is always expressed phenotypically in the organism bearing it, it is called the dominant allele (*A*); a recessive allele, expressed only when the dominant allele is absent, is designated *a*. A dominant trait is expressed when one or two dominant alleles are present (*AA* or *Aa*) but one can see only the recessive phenotype when both recessives (*aa*) are present. Some alleles are not expressed as either dominant or recessive but are intermediate, and they produce a phenotype somewhere between the phenotypes of the parents (red, pink, white).

Suppose that two individuals differing in genes at only three different loci (genetically active sites on chromosomes) are crossed: *AABBCC* x *aabbcc*. The sex cells (gametes) are carrying *ABC* or *abc*, whereas the first filial generation ($F_{1)}$ or offspring, is entirely *AaBbCc*. Each F_1 offspring can produce eight different types of gametes: *ABC, ABc, AbC, aBC, Abc, aBc, abC,* and *abc* the number of possible genotypes when three genes are involved. For an idea of what occurs when more than three gemes are involved. Inasmuch as each chromosome has many thousands of genes the number of possible combinations is astronomical.

The word chromosome comes from the Greek *chroma*, which means colour, and *soma*—body, because these gene carrying units can be routinely observed microscopically when stained with basic dyes. Genes are arranged in linear order on these chromosomes, which are found in the nucleus of eucaryotic cells.

G (3/4) → W (3/4) = 9/16 GW (yellow, round)
G (3/4) → W (1/4) = 3/16 Gw (yellow, wrinkled)
g (1/4) → W (3/4) = 3/16 gW (green, round)
g (1/4) → W (1/4) = 1/16 gw (green, wrinkled)

Fig. A Tri hybrid Cross

Organisms are classified as eucaryotes if their genetic material is contained in the nucleus of their cells. This includes all life above the level of bacteria and blue green algae. The latter—called procaryotes—are considered to be earlier stages in the evolution of life on this earth. Procaryotes have no nuclei and their genetic material is distributed throughout their cells. In eucaryotes, each individual of a species has a characteristic chromosome number—46 in humans, 40 in the house mouse, 20 in corn, 48 in the potato,and 8 in the fruit fly, (*Drosophila melanogaster*).

Sexually reproducing individuals often have two different alleles (are heterozygous) for many of their genes. Thus, the possible combinations of parental genotypes greatly outnumber the actually ever realized genotypes.

The probability of siblings acquiring the same genetic endowment from both or even one of their parents is infinitely small. This probability decreases as more and more pairs of genes are involved. So, each and every human individual (except for identical twins) is and has been genetically unique.

GENOTYPE AND PHENOTYPE

The observable physical and Behavioural aspects of an organism constitute its *phenotype*. The phenotype is the joint product of the genes that an organism possesses (its genotype) plus the effects of the environments in which it developed and now lives.

It has become increasingly obvious with the predominance of the Behaviourist school of psychology throughout the 1950s that the environment is of particular significance in the development of Behavioural phenotypes. A prime example of how Behaviour may be altered via the environment is phenylketonuria a genetic disease, arising from an inborn error of metabolism.

The enzyme that normally breaks down phenylalanine an essential amino acid—is lowered or missing due to gene malfunction. Unmetabolized phenylalanine accumulates in the brain, and one of the effects is a lowered intelligence quotient (IQ). If the metabolic error is circumvented by a special phenylalanine free diet, IQ is ostensibly improved.

CHROMOSOMAL GENETICS

It might be best to consider the nature of the genotype at this point so that genetic principles are understood before we consider the complexities of the effects due to environmental fluctuations. As noted before, the physical bases of heredity are the genes that are borne on the chromosomes carried in cell nuclei. These chromosomes can be observed during mitosis (when the cell divides and reproduces an exact replica of its genetic content, and in meiosis (including reduction division), where the chromosome number is halved to produce the gametes (spermatozoan or ovum) involved in sexual reproduction.

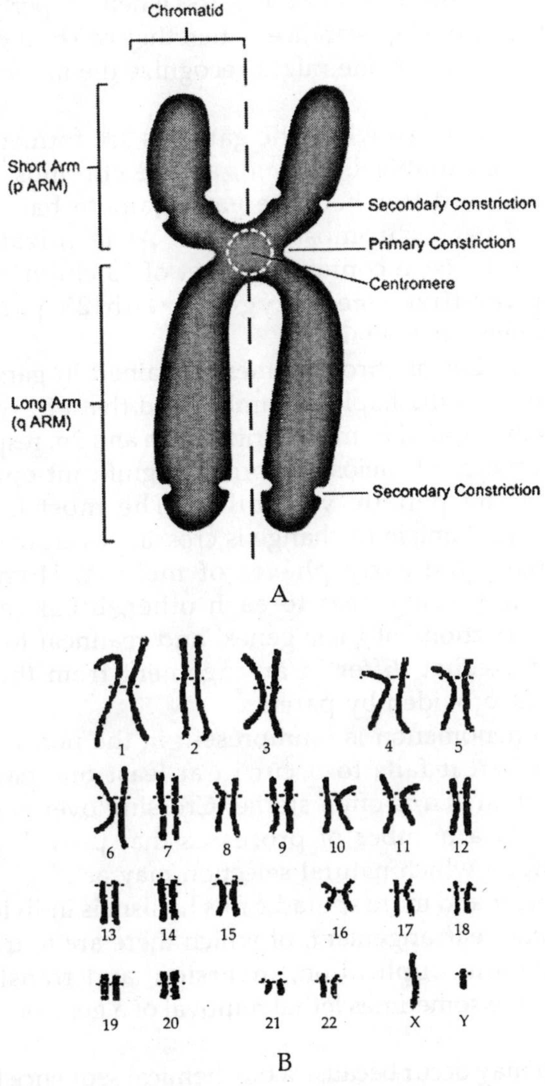

Fig. Human Chromosomes

The human species' 46 chromosomes are arranged in 23 pairs; in females all 23 pairs are homologous (correspond structurally and derivatively); in males, there are 22 homologous pairs plus a nonhomologous *X* and *Y* chromosome. The chromosome pairs, when properly stained, differ in length and appearance so that they can be individually recognized, much as one might recognize the familiar face of a friend.

During meiosis, when the gametes are formed and the chromosome number is halved, all the chromosomes in a single gamete differ because each gamete has only one member of each chromosome pair. At fertilization, two gametes unite, each containing a set of 23 chromosomes to form the fertilized cell (zygote) with 23 pairs or 46 chromosomes once again.

The number of chromosomes contained in gametes (23) is referred to as the haploid number and that of zygotes ($2 \times 23 = 46$) is the diploid number written as n and 2n, respectively.

The process of meiosis affords a significant opportunity for increasing genetic variability. The most important consistent mechanism of change is crossing over, which takes place during the early phases of meiosis. Homologous chromosomes, lying next to each other, break, exchange equivalent portions of their genes, and reanneal to produce combinations that differ in arrangement from the genetic endowment provided by parents.

This phenomenon is omnipresent in the normal meiotic process; when it fails to occur in at least one parent, the resulting organism is often sterile. Crossing over is the most consistent of a number of processes that provide the raw material upon which natural selection may act.

A coarser and more sporadic mechanism is individualized chromosome rearrangement, of which there are four possible types: deletion, duplication, inversion, and translocation. Deletion is the sometimes lethal removal of a gene or sequence of genes.

Death may occur because a biochemical sequence has been excised, resulting in an intolerable structural or functional gap in the organism. Duplication of a chromosome or part of a chromosome may cause an imbalance of gene activity, reducing the viability of an organism. However, because some organisms can tolerate duplications of genetic material, these duplications might play an evolutionary role if part or all of one of the duplications mutated and functioned differently from the original.

Indeed, this is one way in which evolutionary change is postulated to occur. Greater genetic variability gives rise (in minute quanta, to be sure) to physiologic and morphologic differences that endow the organism with an enlarged Behavioural repertoire so that it can more successfully cope with an ever changing environment.

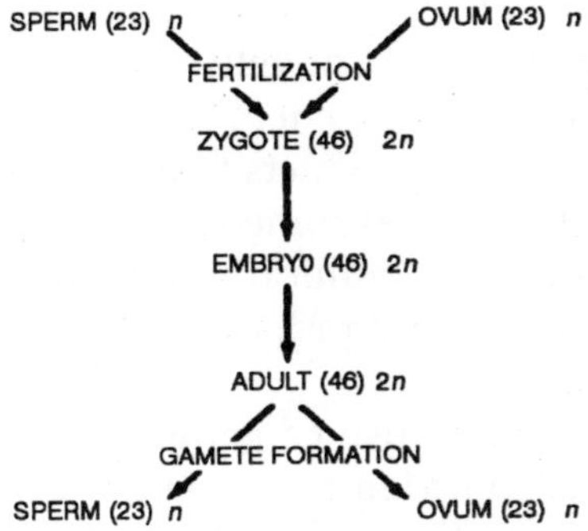

Fig. Human Chromosome Number Changes During Gamete and Zygote Formation

This can be generalized for sexually reproducing organisms, which will be designated by *n*, where *n* is the haploid number and 2*n* the diploid number.

An inversion occurs when a chromosome breaks in one or two places and the segment between the break rotates 180° leading to a reversal of gene order and a resulting change in biochemical product. A translocation occurs when two chromosomes from different pairs break simultaneously and exchange segments.

According to current knowledge, not all these chromosomal and gene changes are equally important in their effects on Behaviour. To date, inversions and chromosome number changes appear to form the main categories of importance. But, crucial to the meiotic process (which reproductive systems evolved to support) is the shuffling and mingling of genetic material into infinite numbers of combinations.

MUTATION AND MUTANTS

Until now this discussion has centered on the shuffling of genetic material in sexually reproducing species. Genetic variability within a species has been discussed with emphasis on the variety of gene combinations attributable to meiosis, crossing over, duplication, and translocation. All these simply involve the reassortment of a set of variable genes indigenous to a population.

The essence of creating novelty is the process of mutation. These are alterations in the chemical structure of a gene that lead to changes in gene products that may ramify throughout the organism. Mutations at some loci occur spontaneously on the average of one in a million newly arisen gametes. The frequency is increased by radiation and by some chemicals known as mutagens.

A veritable storehouse of genetic mutations in mice is found at the Jackson Memorial Laboratory in Bar Harbor, Maine. Through selective inbreeding, when a mutant trait of interest is found, a strain of mice each of whose members possess the trait is established and bred for research uses. Many of the known mutant genes affect the nervous system during its development and produce a range of deficits in Behaviour from minor disturbances to severe defects. One of these found in the mouse is a mutation on chromosome five, an altered gene called *fidget*.

When this gene is present on both members of the fifth chromosome, it produces a Behavioural phenotype in which the mouse's head moves from side to side continuously. The gene is written *fi* and the mouse is said to have an *fift* genotype. In most mice the fidget gene is not present at this locus but the normal alternative gene, written as +, is. If the mouse is genotypically either *fi+* or ++, the Behavioural alteration is not seen. Only if there are two *fi* genes present does the fidget phenotype become visible, as the wild type (+) gene is dominant to the recessive *fi*).

However, dominance and recessivity are not necessarily complete, because heterozygotes (individuals with non-identical alleles) are often distinguishable from both homozygotes (individuals with identical alleles). Even if at first sight dominance appears to be complete, detailed biochemical tests may reveal differences between the heterozygote and the homozygote.

Untreated phenylketonuria in man provides one such example. Superficially viewed, the disease is controlled by a recessive gene *p*. with afflicted individuals being *pp* and normals *p*+ or ++. However, at the biochemical level, *p*+ and ++ may be distinguishable because the *p*+ individual tends to have more phenylalanine in blood serum. Dominance, in this case is incomplete. So, depending on which phenotype component is being examined, different conclusions as to the dominance may be drawn.

If a simple dominant—recessive relationship is assumed with two pairs of alleles at two loci (i.e., *A* and *B* are dominant to *a* and *b*), what might ensue? If a double heterozygote (*AaBb*) is crossed with a double homozygote (*aabb*) what are the statistical expectations? The double recessive homozygote produces only *ab* gametes. From the double heterozygote each separate locus yields equal numbers of *AB*, *Ab*, *aB* and *ab* gametes according to Mendel's principles of segregation and of independent assortment of alleles during gamete formation. If these gametes are fertilized with the recessive *ab* gametes, the results, assuming two independent pairs of alleles and simple dominant recessive relationships, will yield a 1:1:1:1 phenotype ratio.

The backcross to a recessive parent (one with an *aabb* genotype) is often used as a testcross in practical breeding programs because it can determine the actual genotype of an organism whose recessive alleles may be obscured by dominant ones. The use of this wholly recessive dihybrid parent unmasks recessive alleles in the alternate parental genotype. If we return to Mendel's garden peas and specify physical traits in a dihybrid cross (*A* is the gene for yellow and *a* the gene for green; *B* is the gene for a smooth surface and *b* the gene for a wrinkled one), we emerge with four types of gametes (*AB, Ab, aB, ab*) produced in approximately equal numbers, which yield the basic Mendelian ratio of 9:3:3:1.

RECOMBINATION AND GENETIC MAPS

There is an important exception to Mendel's second law. Assortment of genes is not generally independent if gene loci are on the same chromosome. The closer genes are to each other along a given chromosome, the more closely linked they are during gamete formation. Although genes on the same chromosome may recombine with each other during meiosis, the frequency of the recombination depends on the spatial distance between the relevant loci. Generally, the further apart the loci are, the greater the possibility of recombination.

From the frequency of these recombinational events, chromosome maps have been constructed for individual chromosomes. Genes located on the same chromosome are said to belong to the same linkage group. Thus, in humans we have 23 linkage groups (one for each pair of chromosomes, i.e., the haploid number of chromosomes [*n*] corresponds to the number of linkage groups).

The house mouse, possessing 40 chromosomes, has 20 linkage groups; in *Drosophila melanogaster*, an organism of historical importance in Behaviour genetics, the numbers are 8 and 4 respectively. During the formation of human eggs and sperm (gametes), the chromosome number is halved (meiosis), leaving the egg with 22 autosomes and an X chromosome, and the sperm with 22 autosomes and an X or a Y sex chromosome. Males produce four sperm from each specialized cell in the testes and X- and Y-bearing sperm occur in about equal numbers.

GAMETES AND FERTILIZATION

In ovaries, meiosis also produces four products, but three of these become polar bodies, whereas the fourth becomes the chromosome containing nucleus of the developing egg. Human females are born with their entire reproductive quotient already present and the limits are set for the number of eggs they can produce (about 400 during a reproductive life span). Males, however, continue to produce sperm from sexual maturation until senescence.

Differences in gametic size and function are extreme. The head of the sperm is about 3 to 5 microns long and 2 to 3 microns wide. The tail, which propels it, is 10 times the length of the head. Essentially, it is a pared down packet of miraculously organized DNA (genetic material) with a launching and traveling apparatus to respond to and fertilize ova.

The human egg, in contrast, is one of the largest cells in the body—a sphere about 130 microns in diameter. It moves only passively when it is essentially shoved along, taking a few days to progress from ovary to uterus. As noted by Hartl (1977, cited in Probber and Ehrman,: "All the sperm that gave rise to all the people that ever lived could be carried in a teaspoon the eggs, would require a small bucket." The egg is comparatively enormous because in addition to its complement of chromosomes, its cytoplasm provides the nourishment for, and biochemical assistance to the developing embryo for translation of the information carried in its genes.

Unlike the female gamete the function of the sperm as a cell is complete when the sperm contacts and penetrates the egg (having shed its tail), and sperm and egg fuse. With fertilization of the egg, formation of the embryo begins, but it is not until 6 weeks afterward that sex chromosomes become operative.

At this point the *Y* chromosome, if present, will cause the inner layer of the gonads to develop into testes. If, however, there are two X chromosomes, at about 12 weeks the outer layer of the gonads will elaborate into ovaries.

X-CHROMOSOME INACTIVATION

A further complication relates to sex. Although every human embryo is predisposed to develop as a female, its eventual sex phenotype will be determined by the types of chromosomes borne by the sperm. Our species has 23 pairs of chromosomes, each of which carries thousands of genes that compose the genotype that develops and unfolds to become the visible organism, the phenotype.

In humans, 22 of the total number of chromosomes are autosomes; that is, two copies of these chromosomes are present in both sexes. The 23rd pair is made up of the unique sex chromosomes; the female has two of the X chromosomes; the male has one X and one Y chromosome. These were originally called X from X the unknown factor, and Y because it follows X alphabetically.

The X chromosome was first described in 1891 from insect material as "a peculiar chromatin body" but its function was not known then or for a long time afterward. Although the female had two doses of all the genes on the X chromosome and the male only one, there seemed to be no significant differences in X chromosome activity between the sexes. It was not until 1962 that a basis for dosage compensation emerged. This has been named the Lyon hypothesis after Mary F. Lyon, who first presented detailed studies on the subject. She hypothesized that in the normal female, one X chromosome is inactivated in each somatic body cell early in embryonic development. Whether the maternal or paternal X is rendered inert is a matter of chance and the result is that patches of cells with X-chromosomal lineages emerge.

The deactivated X becomes a tightly packed mass usually found close to the inner surface of the nuclear membrane of somatic cells in normal female mammals. It is called a Barr body after its discoverer. Accordingly, each cell in a male or a female has only one operative X chromosome. Because a female's cells are of two types (maternal X and paternal X), she is a genetic mosaic. A male, however, has only his maternal X chromosome to express.

What are the implications of this double X-chromosome dose for the female of our species? At the molecular level the X chromosome is a much more information packed and therefore biochemically active chromosome than the Y, which is much smaller and, except for male sex determination and fertility, inert. Recent work by McKusick and Ruddle cited more than 100 genes in the human X-chromosome map. The Y chromosome bears only two known genes, one controlling the TDF (Testis Determining Factor) and the other coding for the H—Y antigen.

The H—Y antigen plays an important role in the expression of male physiological and morphological characteristics, specifically for the differentiation of the testis. "Further male differentiation is imposed on the embryo by the action of testicular hormones against the inherent tendency toward the female phenotype." It is also known that males with their X and Y chromosomal complement are relatively unprotected from the expression of unpaired deleterious genes on their single X chromosome.

Every thing on their maternal X chromosome is expressed if not modified by genes on the autosomes. This is demonstrated by the much greater percentage of males afflicted by X-linked defects such as colour blindness and hemophilia. Genes on the X chromosome are said to be sex linked. In females the principles of heterozygosity and homozygosity apply to genes on the sex chromosomes just as they do for the genes on autosomes. In males, loci on the X are not matched by loci on the nearly inert Y chromosome; this condition is known as hemizygosity.

EPIGENESIS

The process of development involves constant interaction and coaction of an organism's genes and its environment. Given identical genotypes as in monozygotic human twins or highly inbred mice and *Drosophila,* phenotypes are usually very similar. Even under optimal circumstances however, phenotypic variation occurs because of perturbations in the environment. In a given situation one can speak of the average phenotype of individuals with the same genotype as a *norm of reaction*.

The *range of reaction* covers the spread of phenotypes that could potentially develop from a given fertilized ovum.

Clearly, the range of reaction increases if the environment is not carefully controlled. The *epigenetic* view of development postulates that genotype and environment are involved at every stage from fertilization to birth and beyond. At fertilization, these actions begin at a molecular level.

Enzyme activities and changes in the intracellular environment are initiated. The egg provides a source of energy for the newly formed zygote and a well primed machinery for protein synthesis, which initiates the primary series of cleavages and invaginations that produce the embryo, fetus, and finally, the neonate.

The control mechanisms fostering maturation are encoded within the genes, in the internal and external environment of the DNA, and in the interrelations and interactions between them. At every level molecular, cellular, and organismic there are functional gene environment interactions and products.

Role of the Biosphere

But what is the raw material that comprises the stuff of life that initiates and directs the master plan for the architecture of all the organisms with which we share or have shared this planet? It is a flexible thread that binds the most minute of submicroscopic flora and fauna to the most magnificent (and threatened) product par excellence of evolution—*Homo sapiens*. How did genetic material, those miracle molecules, originate, organize, and evolve?

The condition of primitive Earth (about 4 billion years ago) provided an atmosphere rich in hydrogen, methane, ammonia, and water. In 1952 Harold Urey and and his associates posited the first steps in the evolution of life from inorganic compounds by nonbiologic means. Possible energy sources for the synthesis of these molecules might have been ultraviolet radiation from the sun, lightning, radioactive decay in the earth's crust, and heat from volcanoes and hot springs. The primeval organic "soup".

Haldane was probably present on primitive Earth. As the earth evolved chemically, the atmosphere changed. Most of its hydrogen was lost to outer space and the released oxygen thus gained was converted to ozone and formed above the earth a layer that absorbed ultraviolet light.

After the formation of this layer, the main sources of energy were visible light and heat from the sun.As more synthesis of these primary molecules occurred especially in the oceans more and more intermediate products accumulated in the thickening organic soup. An ingredient of the soup was adenine one of DNA's bases.

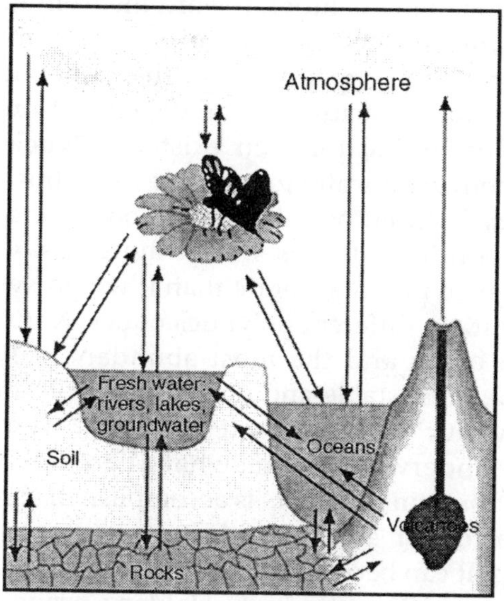

Fig. Major Parts of the Biosphere

It was plentiful in that broth because it was easily synthesized; a simple compound of hydrogen cyanide. The other bases cytosine, guanine, and thymine are more complex structures. Irradiation of solutions of adenine, ribose, and phosphate compounds with ultraviolet light (2400-2900 Å) has led to the synthesis of AMP (adenosine mono-phosphate), ADP (adenosine diphosphate) and ATP (adenosine triphosphate) the basic energy currency of cells.

This means that these high energy phosphate compounds would have been part of the "soup" when life initially evolved and that these adenosine compounds would have been the most common. It is interesting to note three steps that are a precondition for the formation of DNA:

- The union of base and sugar to form the nucleoside.
- The union of nucleoside and phosphoric acid to form the nucleotide.
- The union of nucleotides to form DNA (nucleic acid).

Finally, the union of amino acids to form polypeptides is accomplished simply by the removal of water.

The next step in evolution toward life from monomers to polymers (and collections of polymers) came with the development of interdependence of these chemical systems. Proteins and nucleic acids became dependent on each other in their evolution and for their existence. Nucleic acid can function nonenzymat ically to replicate itself; this is known as autocatalysis. Such autocatalytic function means that a pool of raw nucleotide materials would more easily synthesize copies of a given polynucleotide than it would synthesize, by complementation, different polynucleotides. Natural selection would then occur and the most abundant polynucleotide would be the most stable and quickest to replicate.

With the nucleic acids, proteinoids that were present on early Earth underwent a somewhat altered evolution. The formation of protein polymers is accomplished simply, but the polymerization of amino acids to protein chains is not spontaneous. It can become so if the concentrations of reacting substances are sufficiently high. Haldane proposed that this was accomplished by the concentration of amino acids in drying pools of water near seashores.

Nearly all the 20 amino acids common to proteins were included in these proteinoids, which were very similar to currently known proteins and polypeptides of corresponding size and all showed at least some catalytic activity. Although natural proteins are catalysts, unlike nucleic acids they are poor autocatalysts, so the products and diversity of protein catalysis are determined more by the amount and type of reactants involved than by the qualities of the catalyst. In 1965, Fox discovered a distinct propensity on the part of proteinoids to form microspheres about 2 microns in diameter when hot concentrated solutions of them are slowly cooled. These microspheres show a double layer boundary resembing a membrane (but without lipids) and swell or shrink as ambient salt concentrations vary. They are not alive or are not even direct precursors of living cells, but they demonstrate the effects of forces that would have operated on primitive living systems as they evolved.

If allowed to stand for several weeks these microspheres absorb more proteinoid material from solution, produce buds, and sometimes divide to produce second generation microspheres. This is not the ancestral paradigm of reproduction. It simply illustrates how one of the functions usually associated with living organisms can be mimicked by the remarkable self organizing properties of simpler chemical systems. Thus, it appears at least possible to speculate that the immediate precursors of life were capsules of chemical reactions called *coacervate drops* by Oparin.

He saw them as a model for the evolutionary development of cell membranes, as coacervate preparations form droplets with an acqueous solution surrounding them. These drops will form in many kinds of solutions of polymers including proteins or nucleic acids. Although this droplet forming property is also a function of purely chemical systems, the Behaviour of these sorts of molecules sheds light on the origin of life. More stable coacervates could survive longer and perhaps grow by adsorbing the remains of less stable droplets.

As a result of selection, accelerated rates of reaction would become advantageous and more and more efficient and elaborate catalysts would be developed and retained until evolution of such a system produced the ultimate catalyst—the organic enzyme. This led to what Oparin described as protobionts—organized metabolizing systems. These protobionts delineated by Oparin lacked one crucial quality—they were not capable of reproduction; they did not have the capacity to replicate themselves and to persevere as a species. The mechanism for doing this, the molecules of instruction for self duplication, were still evolving in the polynucleotides of DNA.

BIOCHEMICAL GENETICS

DNA (known fully as deoxyribonucleic acid) is present in almost all cells of a given organism in the same amount and type. The genes (which are mostly organized in chromosomes) are segments of a DNA molecule that specify the structure of a single type of protein. Although genes ultimately determine all the characteristics of a cell, they do not do so directly. DNA exerts its control over a cell by synthesizing the closely related ribonucleic acid—RNA—which in turn directs the synthesis of proteins. Experiments with microorganisms clearly show that DNA contains all information necessary to make daughter cells that are essentially identical to parent cells. The amount of DNA per cell is not large; in the human cell there is about 6×10^{-12} g. In spite of this minute quantity, the amount of information contained is enormous—sufficient to direct the synthesis of a human individual. A DNA molecule is made up of chemical units called nucleotides, each consisting of:

- A base that is a purine [adenine (A) or guanine (G)] or a pyrimidine [cytosine (C) or thymine (T)].
- A pentose (5-carbon sugar), deoxyribose.
- A phosphate group.

In a given DNA molecule, the phosphate and sugar groups are identical and only the bases vary. The information determining heredity resides in the arrangement of the bases, A, G, C, and T. The quantity of these bases per cell is constant within a given species but varies among species. However in every species, the quantity of *A* = T and that of G = C, proving that *A* is always paired with T and G with C.

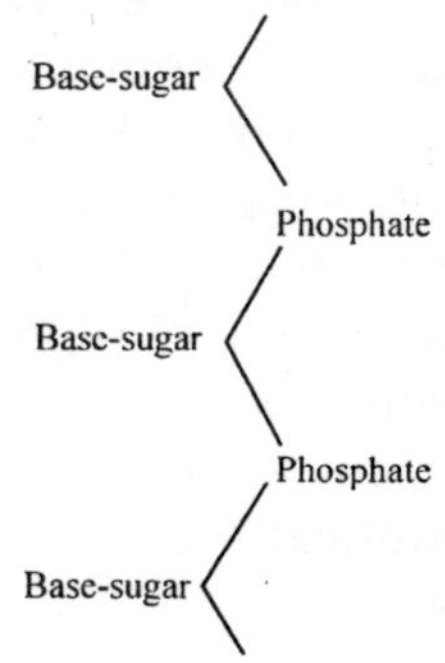

DNA was chemically kn own in the 1800s but the first comprehensive elaboration of the structure and coding of genes was demonstrated in 1953 by Watson and Crick. The bases are attached to a sugar—phosphate backbone forming a chain of nucleotides:

Watson and Crick found DNA to be a double chain of nucleotides:

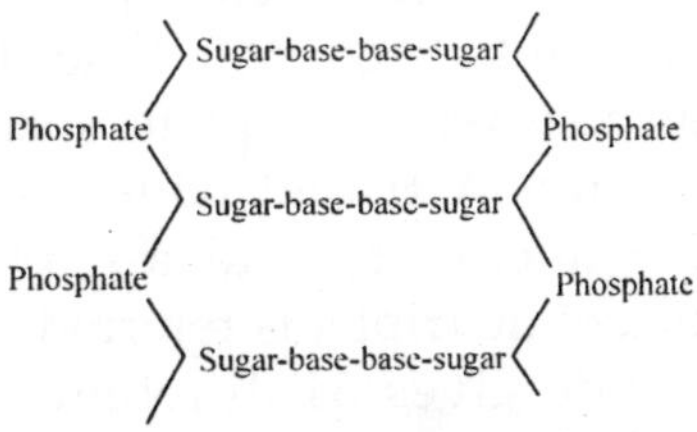

The two chains are joined by the bases, pairing by the rule stated previously and are twisted around each other in a double helix). The nucleotide pairs are separated by 3.4 Angstrom units (Å; an Angstrom unit equals 10^{-7}mm), and the whole structure repeats itself in 10 pairs at intervals of 34 Å. Thus, by applying the pairing rules (A=T and G=C), if we know the sequence of bases on one helix, we also know the sequence on the other.

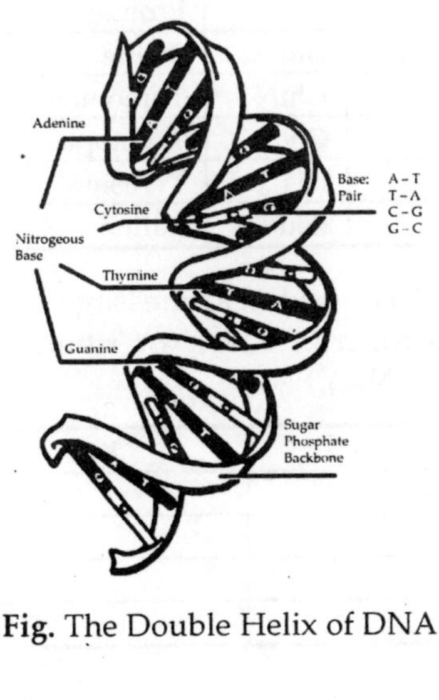

Fig. The Double Helix of DNA

According to the Watson Crick theory, the linear sequence of nucleotides is characteristic for a given species although there are minor heritable variations that lead to variations in proteins that have as their primary structure a chain of amino acids. Therefore, the linear sequence of nucleotides can be regarded as a *genetic code* specifying the proteins that are essential components of all living organisms.

There are 20 essential amino acids specified by the genetic code. Because there are four possible bases (A, T, G, and C), one and two letter codes are inadequate, specifying 4 and 4^2 = 16 combinations only. A triplet code specifying 4^3 = 64 combinations is a necessary and apparently sufficient minimum. The nucleotide triplet is referred to as a *codon,* and because a triplet code gives 64 different combinations, or words, of which only 20 are needed for amino acid synthesis, the code is referred to as a *degenerate code.* In fact some amino acids are coded by more than one triplet. The names and abbreviations of the 20 essential amino acids are:

Alanine	Ala	Leucine	Leu
Arginine	Arg	Lysine	Lys
Asparagine	AspN	Methionine	Met
Aspartic acid	Asp	Phenylalanine	Phe
Cysteine	Cys	Proline	Pro
Glutamic acid	Glu	Serine	Ser
Glutamine	GluN	Threonine	Thr
Glycine	Gly	Tryptophan	Tryp
Histidine	His	Tyrosine	Tyr
Isoleucine	Ileu	Valine	Val

From the DNA code a linear message of triplets or codons is *transcribed* to a form of RNA (or ribonucleic acid) called *messenger* RNA (*m*RNA).

Base in DNA	*Base in mRNA*
A	U
T	A
C	G
G	C

RNA is chemically very similar to DNA except that:

- Its sugar is ribose rather than deoxyribose;
- It has the base uracil (U) substituting for thymine;
- It is single rather than double stranded. In *transcription* from DNA to *m*RNA, the following pairing rules therefore occur:

There are three kinds of RNA on which protein synthesis depends: messenger RNA, transfer RNA, and ribosomal RNA, all carrying codons that correspond to DNA.

First position (5' end)	U	C	A	G	Third position (3' end)
U	UUU, UUC: Pho UUA, UUG: Leu	UCU, UCC, UCA, UCG: Ser	UAU, UAC: Tyr UAA Slop UAG Slop	UGU, UGC: Cys UGA Slop UGG Trp	U C A G
C	CUU, CUC, CUA, CUG: Leu	CCU, CCC, CCA, CCG: Pro	CAU, CAC: His CAA, CAG: Gin	CGU, CGC, CGA, CGG: Arg	U C A G
A	AUU, AUC, AUA: Ike AUG	ACU, ACC, ACA, ACG: Thr	AAU, AAC: Asn AAA, AAG: Lys	AGU, AGC: Ser AGA, AGG: Arg	U C A G
G	GUU, GUC, GUA, GUG: Val	GCU, GCC, GCA, GCG: Ala	GAU, GAC: Asp GAA, GAG: Glu	GGU, GGC, GGA, GGG: Gly	U C A G

Amino acid names:

Ala = alamine	Gln = glutamine	Lou = loucine	Ser = serine
Arg = arginine	Glu = glutamale	Lys = lysine	Thr = threcnine
Asn = asparagine	Gly = glycinc	Mel = methicnine	Trp = threcnine
Asp = aspartake	His = histidine	Phe = phenylalanine	Trp = tryplophan
Cys = cysteina	lie = solvcine	Pro = proline	Tyr = Tyrosine
			Val = valine

Fig. The Genetic Code

Before amino acids are assembled into protein chains they must be "activated" by the attachment of a special phosphoric acid group. They are then attached to another type of RNA called *transfer* RNA (*t*RNA). In fact, there are as many varieties of *t*RNA molecules as there are triplets determining amino acids. Alignment of the *t*RNA and *m*RNA so as to assemble protein chains in an orderly way is mediated by special particles in the cytoplasm of the cell called *ribosomes,* which are made from a third form of RNA called *ribosomal* RNA (rRNA).

The important point then is that the amino acid sequence in proteins is directly determined by the genetic code carried by the DNA molecules even though the mechanism by which this occurs is complex.

A great deal of additional information about this process can be found in many texts, but the details of the process, largely established in microorganisms, are only of preliminary interest to Behaviour geneticists at the present stage in the development of our field. On the other hand, with time there will certainly be a trend toward metabolic explanations of Behavioural processes, so that an understanding of the biochemical basis of gene actions will assume progressively more importance.

Regulatory Processes

So far as protein synthesis itself is concerned, most proteins are made only when they are useful. Therefore, there must be a means of *regulation.* In fact, *regulator genes* have been discovered in microorganisms. These regulator genes determine whether the genes that specify types of proteins (*structural genes*) are active or otherwise.

The regulator genes themselves are controlled by events taking place in the cytoplasm of the cell and thus are open to environmental influences; for example, if a certain amino acid is necessary for growth and is present in the environment, then the cells can cease making the amino acid as well as the enzymes necessary for its synthesis (enzyme repression).

Undoubtedly this process of regulation of protein synthesis must be the basis of *differentiation,* which is the development of specialized cells and tissues. At all stages of development, it is clear that different portions of DNA are active in each type of cell and tissue. Such temporal regulation of gene action must be carefully studied in order to understand Behaviour.

It is clear that genes act in sequence during development so that a given gene may start one event which in turn activates other genes, leading to a series of developmental steps. Gene—hormone interactions, for example, are probably involved in sex differentiation, the onset of puberty, and the development of learning in man. The other important consequence of a knowledge of the physiological processes of a gene substitution is that their pattern can be modified by appropriate treatment. A good example of environmental influence on genes in man is phenylketonuria, which we have already considered in other respects.

Their skin and hair pigmentation is generally lighter than that of the population from which they arise. Phenylketonuria is due to a deficiency or absence of the enzyme phenylalanine hydroxylase, which is necessary in the metabolism of phenylalanine, an amino acid that is an essential dietary constituent.

Normally, phenylalanine → tyrosine →... → various metabolic breakdown products. In phenylketonuria, this first step is blocked; phenylalanine accumulates to a level 40 to 50 times that found in unaffected individuals and this excess leads to mental deficiency. Hence, a likely treatment would be to feed a phenylalanine free diet.

This poses problems, as all known proteins contain phenylalanine. However, such a diet can be synthesized by breaking down protein and reconstituting it without phenylalanine, but still containing other essential amino acids. The diet must begin early in life to have a major effect on IQ and is likely to be proportionately less effective if delayed beyond infancy.

The treatment must strike a delicate balance between malnutrition (i.e., inadequacy of the essential amino acid, phenylalanine) and intoxication. More recently management of phenylalanine levels by regulating enzyme levels in the blood has been successfully achieved and may lead to even more effective control of this once irreversible and untreatable genetic disease.

QUANTITATIVE GENETICS

The focus of classical Mendelian genetics concerned individual traits as reflected in the phenotypic characteristics (colour, structural, and chemical differences) that resulted from the genotype—a product of segregating particulate units that assorted independently. Inferences concerning the gene were derived from data produced from breeding experiments. The distribution of phenotypes was seen as discrete (i.e., falling into distinct classes such as red, white, or pink, according to the dominance or recessiveness of specific alleles).

Genetic variation was conceived of as being under the control of specific genes assignable to specific loci on chromosomes. This approach is often not applicable to traits that are quantitative and do not segregate into discrete classes. An example of this in humans is IQ within a population. This does not mean that specific genes affecting these traits are not present, only that they cannot be precisely identified with techniques currently available. There is evidence, however, that IQ is influenced by genes at known loci and by specific chromosomal anomalies.

Quantitive genetics is the study of variation in those traits where genes responsible for variability in phenotype cannot routinely be individually recognized. This type of inheritance, leading to continuous variation in complex characteristics (such as skin colour, disease resistance, or IQ) is called polygenic because it involves the summation of the action of many genes at different loci that are often cumulative in their effect on a particular trait.

Quantification of genetics began even before the rediscovery of Mendel's work in 1900 by Hugo de Vries, Carl Correns, and Erich von Tschermak. It probably had its origins in the work of Adolphe Quetelet, a Belgian mathematician and sociologist who was best known for his application of statistics and probability theory to social phenomena. He conceived of the idea of the "average man" as a central value about which measurements of a human trait may be grouped according to the normal probability curve.

A further development of mathematical method apropos genetics was provided by Sir Frances Galton—a cousin of Charles Darwin and one of the first to recognize the implica tions of Darwin's theory of evolution for human traits (including cognitive ones).

Although Galton is perhaps best known by Behaviour geneticists for *Hereditary Genius* in which he argued that mental and physical features are both inherited, he concerned himself for more than three decades with the improvement of standards of measurement in a variety of disciplines.

This concern was carried further in the work of his disciple, Karl Pearson, one of the founders of modem statistical science. Influenced by Galton's evolutionary writings, Pearson became enamored with the notion of applying statistics to the biologic problems of heredity and evolution. These efforts in turn led to an assortment of methods that are part of the core of statistics: the chisquare test of statistical significance and the correlation coefficient or Pearson.

For a time this resulted in a rather acrimonious confrontation between the biometricians led by Pearson, who touted the quantitative aspects of heredity, and the Mendelians, inspired by William Bateson founder of the science of genetics who argued for the unit character of inheritance and discrete classes of traits.

Using pure lines of beans that produced large or small plants from beans of corresponding size, he crossed tall with dwarf plants and demonstrated that the mean value of a measureable trait in this case, weight of the beans produced by the progeny of these crosses tended to approach the mean of the total population and not to reflect either of the parental extremes.

Although these experiments provided a basis for the notion of a continuously varying heredity, the accommodation of this theory with the Mendelian concept of transmission of discrete units of inheritance was furnished by the work of a Swedish geneticist, Nilsson Ehle, who discovered a hereditary trait with distinguishable classes the depth of red pigment in the hull of a wheat, *Triticum vulgare.*

This plant exhibited a graded series of redness from deep red to white (with five intermediate shades), which formed a distribution encompassing easily distinguished discontinuous groups. The variation was controlled by three independent loci with pairs of red or nonred alleles at each locus and with each allele for pigment contributing additively to the total phenotype. The documentation of discrete qualitative differences within an otherwise continuous distribution, that of seed colour in wheat, was explained by Nilsson Ehle in terms of a continuous graduated character determined by three Mendelian pairs of alleles with additive nondominant effect; that is, with multiple factors, later called polygenes. These crucial experiments assisted in the reconciliation of the biometricians' blending (continuous variation) hypotheses with the Mendelians' discrete units of heredity.

One of the peripheral issues that emerged from this dispute was the distinction between more easily traced major genes, which have massive effects on the phenotype of organisms, and minor genes with less striking phenotypic effects. In this latter instance, several minor genes (polygenes) may contribute to the formulation of a characteristic. For example, body weight, height, and intelligence in humans are influenced by many genes that act in more or less additive ways.

And although it is the cumulative number of more or less contributory alleles that determines the individual phenotype, it is the arrangement of alleles on specific chromosome pairs that influences genetic transmission of the trait. By the time a gene is expressed, its capacities have been altered by its association with other genes as well as by environmental interactions.

To better understand the nature of continuous variation let us consider, as an example, height. In this instance it is not possible to place people into discrete classes such as tall or short because all degrees of height occur. The frequency distribution of many quantitative traits approximates the *normal distribution*. It can be completely described in terms of two parameters.

One is the mean or average value; the other is an expression of variability around the mean. In some cases the variability around the mean is small (Curve A; elsewhere, the population distribution may be more diffuse, and the variability around the mean is larger (Curve *B*). The term for the quantity measuring variability is *variance,* which is the sum of squares of deviations around the mean divided by $n - 1$ (*where* n = the number of observations). The square root of the variance is the *standard deviation.*

It can be shown that 95per cent of the population of a normally distributed trait lies within two standard deviations on either side of the mean. Much of the theory of quantitative genetics is based on the assumption of a normal distribution. If it is impossible to assume a normal distribution, a suitable algebraic transformation may be found that will convert the data to an approximately normal distribution.

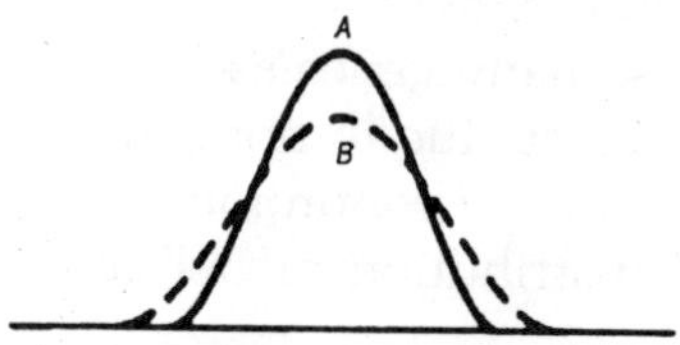

Fig. Normal Distribution Variance of B is Larger than that of A

Components of Phenotypic Variance

If we assume that a trait such as height is partially genetically controlled we must ask how the intrinsically discontinuous variation caused by genetic segregation is translated into the continuous variation of quantitative traits. Suppose two individuals *Aa* and *Bb* are crossed, where *Aa* and *Bb* signify gene pairs at two unlinked loci.

Further, if genes *A* and *B* act to increase the measurement of a quantitative trait by one unit and genes *a* and *b* act to decrease the trait by one unit, we can arrive at a quantitative value for a genotype by counting the number of plus (+) and minus (-) genes.

The cross would yield five genotypes, the most common genotype +/– +/– (*Aa Bb*) having a genotype value of zero; this is the *mean genotypic value.* The least common genotypes are the two extremes +/+ +/+ (AA BB) and –/– –/– (*aa bb*) with values of +4 and –4 respectively. If there is a third locus with two similar alleles, then a cross between multiple heterozygotes would raise the number of genotypic classes to seven, a fourth locus to nine, and so on. As more genes are channeled into the expression of a trait, the more smoothly continuous the distribution appears. This is true because the differences among the classes become progressively smaller as the number of segregating loci rises.

At the stage when the differences among classes become about as small as errors of measurement, the distribution becomes continuous as in. Then too, any variation due to nongenetic causes, (i.e., to environmental influences) blurs the underlying discontinuity of the segregating alleles still further so that the variation is seen as continuous irrespective of the accuracy of the measurement.

Because of this, many genes each with small effects on a trait plus the variability due to nongenetic or environmental causes lead us to expect a continuous distribution of a given trait—the normal distribution or bell shaped curve already cited.

Genes that contribute to a quantitative trait but are not directly identifiable by classic Mendelian segregation (i.e., cannot be studied individually) are polygenes. Genes whose effects can be studied individually are referred to as major genes. The distinction does not imply any fundamental difference between the two categories of genes.

It is merely a matter of convenience of nomenclature because the breeding methods used to study the effects of major genes cannot in general be used in the study of polygenes. Quantitative or biometrical genetics has been developed to deal with the variation caused by the simultaneous segregation of many polygenes.

Heritability

Behavioural traits such as activity scores in mice, duration of copulation in *Drosophila*, or IQ in humans are essentially quantitative. To analyse these traits an appreciation of the aims and methods of quantitative genetics is essential. One of the main goals of biometrical genetics is to make known the apportionment of the total phenotypic variance (V_p) associated with a given trait into components due to the genotype (V_g)

and to the environment (V_e). If we assume no interaction between genotype and environment—the simplest assumption possible—then phenotypic variance is the sum of the genotypic and environmental variances: $V_p = V_g + V_e$. It is reasonable then, to compute the proportion of the total phenotypic variance that is genotypic:

$$\frac{V_g}{V_g + V_e} = \frac{V_g}{V_p}.$$

This ratio is referred to as the *heritability* of the trait *in the broad sense.* However, the important measure in plant and animal breeding is *heritability in the narrow sense.*Heritability varies from zero when the genotype is unimportant to one when the trait's determination is entirely genotypic, as with blood groups. It should be noted here that the estimated heritability refers to the population under analysis and not to the trait itself. A different population may give a different heritability for even the same trait.

There are several ways of calculating heritability. One method is based on the use of laboratory animal strains that have been inbred for many generations (usually via sister x brother matings) and are, by the end of the breeding regime, essentially homozygous.

This means that *within* each of these strains all individuals are "genetically identical," so that any variation that occurs will be due to nongenetic causes. However, *between* strains there will, in general, be variation due to the different strains as well as to environmental variation. Using a statistical technique known as "analysis of variance," (which was devised by R. A. Fisher) the variation within and between strains can be separated. This method enables us to calculate estimates of V_G and V_E and therefore to compute heritability.

The average effect that an allele has, in all genotypes in the population that carries it, is its *additive genetic value*. Deviations in any one individual may be due to dominance effects, epistasis (an interaction where one gene modifies the effect of another at a different locus), and the environment. Certain breeding procedures allow division of the genotypic variance, V_G into variance due to additive genes, V_A, dominance, V_D and epistasis, V_I, so that we can formulate $V_G = V_A + V_D + V_I$. This permits a more refined definition of heritability as the ratio of the additive genetic variance to total variance V_a/V_p, which is referred to as *heritability in the narrow sense* and written as h^2 in contrast to heritability in the broad sense (V_g/V_p), or degree of genetic determination. The parameter h^2 is a measure of the proportion of variation due to additive genes and as such is a more useful concept than degree of genetic determination.

(Heritability in the narrow sense provides a measure of genetic variability transmitted from one generation to the next because it takes into account only additive genes and provides a basis for selective breeding.) It has been widely applied in the study of quantitative Behavioural traits including intelligence, temperament, emotional Behaviour, specific abilities, and some other aspects of personality (e.g., neuroticism). The evidence strongly supports polygenic influences on these traits.

Parents pass on fractions of their genes via gametes to succeeding generations, not their entire genotypes. This means that dominance and epistatic relationships in parental genotype are, in general, broken down and not transmitted from one generation to the next in an outbreeding population. These altered gene interactions have important inter-generational implications for families in search of genetic counseling.

Stabilizing, Directional and Disruptive Selection

There are three primary types of natural and artificial selection to which a quantitative trait may be subjected. The form whereby intermediate phenotypes are favored at the expense of the extremes is referred to as stabilizing selection. For a trait such as height in humans, it seems reasonable to expect a normal distribution, as the extremes of very short and very tall are disadvantageous because they are less likely to reproduce than the median individual and so, on the average, phenotypically extreme individuals will contribute fewer genes to subsequent generations. This type of selection is common in natural populations. Evidence for stabilizing selection has been found for many traits (e.g., for hatchability of duck's eggs of assorted sizes and for birth weight of humans in relation to longevity).

Thus, in natural populations, selection will tend to reduce the variability of many quantitative traits by favoring intermediate phenotypes. Because quantitative traits are under the control of many genes, unique genotypes, and hence extreme phenotypes, will be produced continuously by genetic segregation and recombination but will usually be at a selective disadvantage and are likely to be eliminated through natural selection.

Yet because environments can never be constant, some extreme variants may be advantageous in part of the geographic range of a population or at certain times or both. Further, some of these variations may prove useful if the organism is challenged to adapt to an environment that is continuously changing in one direction. The converse of stabilizing (or canalizing) selection is diversifying (disruptive) *selection,* wherein the two extremes of a distribution are favored.

This can occur when an interbreeding population occupies different ecological niches, perhaps seasonally. In one of these, a particular phenotype may be favored because of specific niche qualities to which it is better adapted, but in the other niche a different phenotype may prevail. The importance of this type of selection is still speculative.

Finally, there is *directional selection,* in which one extreme is favored generation after generation. This is the form of selection employed by many plant and animal breeders favoring traits of economic importance (high yield and disease resistance in wheat, fertility in poultry, size and body weight in cattle, etc.) In Behaviour genetics, one of the main experimental methods often employed involves selection of individuals at the extremes of a distribution and by directional selection, forming separate high or low lines in subsequent generations.

Using special breeding techniques, it is even possible in some cases to localize genes controlling a quantitative trait to specific chromosomes or even to regions of chromosomes. In predicting a response to directional selection, the mean value of the populations selected must be considered. This will depend on the size of the population available at the outset, which, in animals, will be finite.

The likelihood of inbreeding will be increased if the selected population is too small, which will lead to an excess of homozygotes and a reduction of the level of genetic variability. Frequently, fertility and general fitness decrease during selection because these more homozygous genotypes with their lowered variability have not been exposed to the severe action of natural selection. Indeed their response must diminish rapidly over generations.

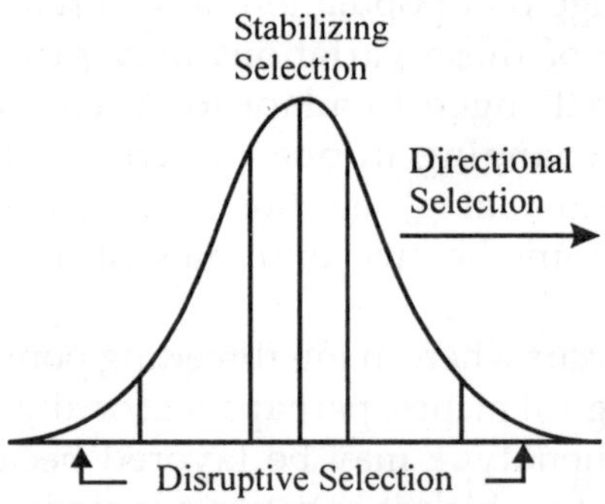

Fig. Directional, Disruptive and Stabilizing Selection

Examples of successful selection for Behavioural traits include ferocity or gentleness in dogs, egg production in chickens, speed and strength in horses, and mating speed in *Drosophila*.

POPULATION GENETICS

Basic to the work of Fisher, Haldane, and Wright was the notion of the mathematical consequences of Mendelian heredity. This is embodied in the Hardy Weinberg equilibrium principle. It resolved the question of what proportions of alleles at a given locus would be propagated in a large mixed population. Indeed, the study of population genetics is based on this principle. Whereas Mendelian genetics is concerned with gene segregation and action in progenies at the familial level, consideration was now extended to a population (a randomly breeding group of individuals).

Much of population genetic theory is based on the assumption of random mating (panmixia), but this does not necessarily always apply. One important modification occurs as a result of inbreeding—the mating of individuals related to each other by ancestry. Individuals with a common ancestor are likely to carry replicas of an allele present in that ancestor and, if they mate, both replicate alleles may be passed on to their offspring. This process increases the proportion of homozygotes compared to that produced by panmixia.

Phenotypic assortative mating is the mating of individuals based on phenotypic resemblance. This occurs when like phenotypes mate more frequently than under random mating. Because like phenotypes may be under the control of like genotypes, this will lead, as will inbreeding, to more homozygotes than expected via panmixia. Positive assortative mating has been found for many traits in humans (e.g., for height, educational attainment, and various Behavioural traits).

The geneticists' infinitely large, randomly interbreeding population constitutes a gene pool the total genetic information of the breeding members of a population of sexually reproducing organisms. The Hardy Weinberg law states that in a large, randomly mating population, gene frequencies and genotype frequencies will remain constant from generation to generation unless outside forces act to change them.

In the "simplest" situation, where the frequency of the alleles *A* and *a* are designated as *p and* q respectively, the basic equilibrium formula is $p^2(AA) + 2pq(Aa) + q^2(aa) = 1$. Such population equilibrium is based on several assumptions: each individual's genotype must be as reproductively fit as that of others in the population; and random (panmictic) mating must consistently occur throughout the population, which must have a large number of individuals, bearing no newly arisen genetic variation (mutation). Essentially it is a description of a static Mendelian population—one not evolving.

Mathematical formulations explaining population changes in allele frequencies were provided by Fisher, Wright, Haldane, and others who conceived the theoretical models and investigated the mechanisms of change that allow the exploration of the genetic architecture of different populations.

Population dynamics are a complex interaction of changes in allele frequency caused by selection, mutation, chance (random genetic drift), and differential migration. Also important are the frequency of chromosomal changes—duplications, deletions, inversions, and translocations and that of numbers (ploidy) which provide the basis for inquiry into the relations of individual genotypes and their ever changing environment.

A by now classic illustration of the changes in allele frequency in a population as a result of a changed environment may be seen in the studies conducted by Kettlewell. He collected elsewhere and released in grimy industrial Birmingham 447 black moths *Biston betularia.* At the same time 137 white moths (also not of the vicinity) were released. After a period of time, moths in the area were caught and classified and the black moths (who blended best with their surroundings) had a 2:1 survival advantage over the white ones. Note that camouflaged insects are obliged to manifest suitable Behaviours (i.e., they must settle on an appropriate background).

The converse of this experiment was performed in nonindustrialized, rural Dorset where the survival rate of the white moths was three times that of the black. The implication is that certain genotypes in a species provide their carriers with fitness traits (in this case, colour) that are advantageous for survival and reproduction, but these can be altered dramatically in value with correlated environmental alterations. And what natural environment is constant?

ORGANIC AND CULTURAL EVOLUTION

Environmental variability is one of the conditions of

organic evolution. Dobzhansky stated that evolutionary theory provides a basic paradigm for the natural sciences: "Nothing in biology makes sense except in the light of evolution" Humans are animal in origin and in biologic characteristics—anatomically similar to the great apes, particularly the chimpanzee and gorilla. Also the chemical structure of our proteins (especially hemoglobin) is very similar in humans and in apes. These similarities suggest that human ancestors were apes along an evolutionary continuum that produced gorillas, chimpanzees, and humans—a sequence that is supported by the fossil record.

But we are much more than merely biologic, as no other species has the capacity to control its destiny as we do, nor the ability to remember and profit from the past, communicate this experience to others of our species and look forward to and plan for the future. These abilities have enabled humans to change their way of life at a dazzling pace and in formidable fashion—faster than any other species and in a qualitatively different way; and this has occurred without our undergoing any major change in anatomical features or brain structure for the last 35,000 years. Thus an understanding of human evolution involves the recognition of both its biologic and its cultural aspects and the interrelations between them.

Though human anatomic traits have not varied much in 35,000 years, human society has changed so rapidly and in so many different ways that our species has established control over the inanimate environment and other animals. What we do with this dominance is the subject of other texts and disciplines (philosophy, social science, ethics, etc.) but it has been gained not by organic but by cultural evolution. The latter resembles organic evolution in that both processes result in changes, often directional, involving both a population and its environment.

Like organic evolution, the cultural variety can proceed in many directions and at differing speeds but the stored variability that makes cultural evolution possible is not genetic in nature. It consist of ideas, inventions, laws, traditions, and all the other learned responses by which society is regulated.

Individuals acquire their share of it not by heredity but through conditioning, training, and learning; and the stored variability of culture increases not genetically, but by additions to and synthesis with the ongoing collection of inventions, ideas, and customs that connote our culture.

Another profound difference between organic and cultural evolution is the way that changes in the relationship between population and environment occur. Alfred Russell Wallace, whose evolutionary theory strongly paralleled that of Darwin, noted that animals evolve by adjusting themselves to their environment; man by adjusting the environment to himself.

These considerations exist on a more restricted level with other species. Bonner notes that: "Many birds and mammals have a primitive form of culture in which there is a transmission of Behavioural information and there is a maintenance of accumulation in the form of tradition. The only difference between their culture and that of man is that the transmission of the information is by imitation rather than by true teaching." He cites as examples the "termiting" of chimpanzees, where a twig is poked into a termite nest opening, revolved, and extracted with a snack of termites adhering to it.

Another celebrated case is the opening of milk bottles by titmice in Britain. A particularly bright and/or hungry bird pecked through the aluminum cap on the milk bottle and this skill spread so rapidly that soon most of the bird population of the British Isles was obtaining its breakfast cream in this fashion.

An equally impressive Behavioural innovation was that of a female macaque who washed the sand from sweet potatoes before eating them. All these creative inventions were passed on to other members of the species and are prime examples of the importance of Behaviour in cultural evolution.

But what of the ubiquitous mode of transmission—organic evolution? Recently, Dobzhansky, Ayala, Stebbins, and Valentine stated that: Organic evolution is a series of partial or complete and irreversible transformations of the genetic composition of populations, based principally upon altered interactions with their environment.

It consists chiefly of adaptive radiations into new environments, adjustments to environmental changes that take place in a particular habitat, and the origin of new ways for exploiting existing habitats. These adaptive changes occasionally give rise to greater complexity of development pattern, of physiological reactions, and of interactions between populations and their environment.

The principle of evolution provides a unifying theme for contemporary biologic sciences but the link that makes the "modern synthesis" of evolutionary theory an invaluable paradigm was the contribution of Charles Darwin in his concept of *natural selection* as one of the mainstreams of thought from which the discipline of Behaviour genetics emerged.

Darwin was the first to recognize that the connection between environmental variability and evolutionary change was provided by this phenomenon. In Darwin's words—from the Introduction to *Origin of Species* : As many more individuals of each species are born than can possibly survive; and as, consequently, there is a frequently recurrent struggle for existence, it follows that any being, if it vary however slightly in any manner profitable to itself, under the complex and sometimes varying conditions of life, will have a better chance of surviving, and thus be *naturally selected.*

From the strong principle of inheritance, any selected variety will tend to propagate its new and modified form. One inference is that through natural selection, some individuals in a population survive and some do not. The notion that populations are culled in this fashion came initially from Thomas Malthus' *Essay on Population,* which Darwin read in 1838 when he was formulating his ideas about natural selection.

Darwin deduced that because population growth in a species is checked by some means, not all individuals are equally susceptible to being eliminated. This idea of the variability of individuals within populations led to the hypothesis that the factors checking the increase in numbers of individuals, within a species, act more strongly on those that are relatively poorly adapted and favour those best fitted to their environments. Individuals thus favored will leave more offspring over generations than their less well adapted conspecifics.

This concept that Darwin recognized and documented—that among diverse individuals in a population some have a higher probability of survival than do others—lacked one fundamental. He was not aware of the nature or mechanisms of hereditary variation.

It was the contribution of Gregor Mendel and the combination of Darwinism and Mendelism that provided the basis for the synthetic theory of evolution. Natural selection favors certain individuals in a population and as a result acts on populations of genes controlling traits; so, slowly the genetic constitution of the population changes in response to selection. As an evolutionary force, natural selection acts at the population rather than at the individual level. Artificial selection, whether for exotic breeds of animals or for increased food yield, acts in the same way.

It should be evident that the Behaviour of individuals within a population (mating patterns, habitat selection, resource utilization) is both a result of selection and of the factors that mediate it.

Behavioural phenotypes are manifestations of the underlying genotypic variability of populations. Those Behavioural traits that are relevant to population continuity need to be isolated and studied under a multiplicity of environments because these Behaviours are among the most important components in evolutionary processes.

Chapter 2

Characteristics of Genetic System

X-CHROMOSOME AS TYPE

The chromosome combinations present in the fertilized egg, and this has brought to light some of the methods by which the chromosomes act on development and characteristics.

We now Examine other effects of altering chromosomes. We shall have to examine the effects of altering all the different classes of chromosomesýÿthe X-chromosome, the Y-chromosome, and the different autosomes.

Each of these three groups (X, Y, and autosomes) gives rise, as we shall see, to a different type of inheritance. The simplest relations, from the experimental point of view, are presented by the X-chromosomes. We therefore deal first with these.

EFFECTS ON ALTERING THE X-CHROMOSOMES

The X-chromosomes take such a course in passing from generation to generation that it is possible to follow the descendants of a particular Xchromosome (that present in the original male parent, for example), knowing in which individuals they are present, in which they are absent. In certain individuals (males, in organisms of Group I) there is but a single X-chromosome instead of a pair.

These relations make it a relatively simple matter to discover the distinctive effects of a particular X-chromosome. The effects of X-chromosomes will therefore be dealt with somewhat fully as a type of chromosomal action. Later the action of the other chromosomes will be taken up.

ABNORMAL DISTRIBUTION OF X-CHROMOSOMES

What will happen if the X-chromosomes by accident become irregularly distributed? We know that sometimes the X-chromosomes are indeed irregularly distributed. Will the sex-linked characteristics continue to follow them; will the defective characteristics show the same irregular distribution as do the X's?

Such cases in great number have been fully observed. It is found that the sex-linked characteristics do indeed follow the X-chromosomes wherever they go. This proves conclusively (if there were any possible doubt in view of the extraordinary course normally followed by such characters) that it is indeed the X-chromosomes on which the characters depend. The matter is one of importance and interest, so that it will be worth while to examine carefully certain typical cases of the result of irregular distribution of X-chromosomes.

Normal Distributions

Sometimes the body of Drosophila is yellow instead of the normal grey; this is a recessive character due to a defect in the X-chromosomes. Suppose that we mate together a female that has a yellow body and a male that has the normal grey programmes. The female has two modified X-chromosomes, which we may represent by XX. The male has its X unmodified; we may represent its cells as XY. Normally in forming the germ cells, the two X's of the female separate into different germ cells, and all the ova receive one of the modified X's. The normal male produces germ cells of two classes, X and Y.

When the germ cell X from the female unites with Y from the male, sons are produced, with the constitution XY; while X from the female with X from the father gives daughters XX. Since the yellow programmes dependent on X is recessive, the daughters XX have the normal grey programmes, like the father.

But the sons XY have only one X, and as this is the defective one, from the mother, the sons are yellow, like the mother. The normal result of such a mating therefore is that all the sons are recessive like the mother, all the daughters dominant, normal, like the father. This is what occurs when the X-chromosomes are distributed to the germ cells in the usual way.

ABNORMAL DISTRIBUTION

In forming germ cells both X's of the female go together to a single ovum, while other ova receive no X. In a race discovered by L. V. Morgan the two X-chromosomes in the female were partly united, so that they almost invariably thus go together to one ovum, leaving other ova without an X.

In this race the X-chromosomes were so modified as to produce the recessive yellow programmes described in the preceding paragraph. We therefore have an opportunity to determine the course of inheritance of the yellow and the grey programmes when the X-chromosomes are thus abnormally distributed.

Represent as before the X-chromosomes that produce the yellow programmes by X. Thus one set of ova receive XX, while another set receive no X. When the germ cells XX united with the germ cells Y from the normal father, there were produced daughters XXY, in which both X's came from the mother, and these daughters were yellow like the mother (since no normal X was present). Sons were produced by the union of ova that contained no X with sperms that carried X; this gave sons XO, with their single X from the normal father, instead of from the mother, as is usually the case. And such sons had the normal grey body programmes of the father, instead of the yellow body programmes of the mother, as happens normally.

Thus with a change in the distribution of the X's there is a corresponding change in the method of inheritance. If the sons receive their X from the recessive mother, as in the normal cases, they are recessive like the mother. But when, through non-disjunction, the sons receive a dominant X from the father, they are dominant like the father.

Similarly, in the normal case the daughters receive a dominant X from the father and are therefore dominant like the father. But in cases of non-disjunction the daughters receive both their JPs from the recessive mother, and are therefore recessive like the mother. In sum, when the X's are normally distributed, such matings give 'criss-cross inheritance'; sons like the mother, daughters like the father. But when the X's are not so distributed, there is no criss-cross inheritance; sons are like the father, daughters like the mother. The dominant and recessive characters follow the respective X-chromosomes, whether these are distributed normally or abnormally.

Such experiments have been repeated many times, and with other sex-linked characters. Always the characteristics follow the distribution of the X-chromosomes, however these are distributed.

It may be concluded with certainty that it is the presence of X-chromosomes of a modified type that causes the appearance of the particular sex-linked characters that are manifested. The rules of distribution of sex-linked characters are the rules of distribution of the X-chromosomes.

DOMINANCE AND RECESSIVENESS IN SEX-LINKED CHARACTERS

As we have seen, and shall see further, many defective conditions are due to defects in certain X-chromosomes. These follow from generation to generation the distribution of these X-chromosomes. Most of these bodily defects are manifested only in individuals in which the defective X- chromosomes are the only kind present; that is, these defects are recessive.

But it is important to observe that in such cases the normal condition of the organism likewise follows the distribution of certain X-chromosomes.

Haemophilia follows certain X- chromosomes. But in the same matings in which this occurs, the healthy condition of the blood follows certain other X- chromosomes. If a defective mother is mated with a normal father, the sons are defective because they receive only the mother's defective X-chromosomes.

But in the same way the daughters are normal because they receive the father's normal X-chromosomes. In the normal individuals the X-chromosome plays a part, just as it does in the defective individuals. In the normal individuals it is healthy and supplies what is required for normal development, while in the defective individuals it fails to supply what is required. In the case of most defects this normal condition is dominant over the recessive condition; that is, when both are present, the normal chromosome is the one that prevails in its effect on the individual. There are defective characters, however, in which the defective condition is dominant or partly dominant.

In these cases, when a defective X and a normal one are present together, the defective one produces its effect, as in the case of bar-eye, described earlier. In many such cases the defect is less marked when a normal chromosome is present as well as a defective one; then the defective condition is said to be partly dominant.

Thus whenever an individual bearing defective X-chromosomes is mated with a normal individual, we have both dominant and recessive characters, the course of which may be followed in the later generations the dominant character being commonly the normal or usual condition. How this works out is shown in our next paragraphs.

TESTS FOR SEX-LINKED INHERITANCE

By observing the distribution of the dominant and recessive characters among the offspring of certain matings, a test for determining whether given characters depend on the X-chromosomes is supplied. The test consists in making what are called reciprocal crosses.

- On the one hand mate a dominant female with a recessive male;
- Also mate a recessive female with a dominant male.

Represent a dominant X-chromosome by a capital X, a recessive X-chromosome by a lower-case letter x. In the group of organisms with which we have been dealing, the female has two X-chromosomes, the male but one. The daughters receive an X from each parent, the sons an X from the mother only.

A dominant female will be represented by XX, a recessive male by xo; similarly a recessive female is xx, a dominant male is XO. The two ma tings and their results will then be represented as follows:

- XX by xo gives X + XO Dom. Mother Rec. Father Dom. Daughters Dom. Sons
- xx by XO gives X + xo Rec. Mother Dom. Father Dom. Daughters Rec. Sons

Thus when the mother is dominant, the father recessive, all the children are dominant, like the mother. When the father is dominant, the mother recessive, the daughters are dominant like the father, the sons recessive like the mother; this is called 'criss-cross inheritance'.

Whenever reciprocal crosses give these results, it is certain that the two diverse characteristics (dominant and recessive) result from differences in the X-chromosomes of the two parents. 'Criss-cross inheritance' is particularly useful in showing at once that we are dealing with sex-linked inheritance.

In every case where such results are produced, further tests show that the two characteristics follow in later generations the two different kinds of X-chromosomes wherever they go. It is mainly by the use of these tests that the many different characteristics dependent on diversities in X-chromosomes have been discovered.

SEX-LINKED INHERITANCE IN GROUP II

Sex-linked inheritance as we have just described it was originally discovered in animals belonging to Group I, in which the females have two X-chromosomes, the males but one (with or without a Y- chromosome).

But when the tests we have just described were applied to birds and to certain other animals, it was discovered that there is a second group (Group II) in which it is the male that has two X-chromosomes, the female but one.

In the common fowl, reciprocal crosses were made between 'barred' fowls (Plymouth Rock) and black fowls (Langshan). Here the 'barred condition was found to be dominant. The results were:

- Barred father by black mother=Barred sons + barred daughters.

- Black father by barred mother = Barred sons + black daughters.

Here the second mating gives 'criss-cross inheritance', showing that we are dealing with sex-linked characters dependent on differences between the X-chromosomes of the parents. But, in the first mating above, all the offspring are like the father instead of like the mother, which latter was the case in the matings. The father in this case has the dominant characteristic (barred), and all the offspring are dominant like the father instead of like the mother.

When the mother is dominant ('barred') the result in the fowls is criss-cross inheritance; while, in the cases before described, criss-cross inheritance occurs when it is the father that is dominant. In fact, in the birds, the males and females simply exchange roles, as compared with their roles in the organisms of Group I.

Since the roles of males and females are interchanged in the two groups, their chromosomal conditions must also be interchanged; that is, since in Group I the female has two X's, and in Group II the male plays the same role as the female of Group I, the male of Group II must have two X's, the female but one. The chromosomal conditions in the two groups are shown in the above tabulation, X signifying dominant, while x signifies recessive, and o signifies the lack of X (whether Y is present or not).

Careful examination shows that this is the only way in which the results in Group II can be produced. Birds and certain moths show inheritance of the kind typical for Group II. In both these it has since been found under the microscope that the males have indeed one more chromosome than the females.

ROLE OF DEVELOPMENT IN X-CHROMOSOMES

A great many characteristics, of various kinds, have been found to be the result of modifications of particular X-chromosomes, and thus to follow in later generations the distribution of the descendants of those chromosomes. Examination of a number of these is desirable, both for their own importance and for the light they throw on the functions of the X-chromosomes.

In man, the following characteristics, among others, dependent on diverse types of X-chromosomes, are known from the fact that they show typical sex-linked inheritance.

Haemophilia: Lack of coagulability in the blood. This results from a serious defect in certain X-chromosomes. The existence of defective X-chromosomes having this result shows that the normal X-chromosomes play a part in supplying something necessary for producing normal blood that coagulates properly.

Programmes Blindness: The fact that defectiveness in X-chromosomes causes programmes blindness shows that the normal X's play a role in producing the normally functioning eyes.

NIGHT BLINDNESS

- Inability to see in a poor light.
- Near-Sightedness of certain types.

Progressive atrophy of the muscles. The fact that defects in X-chromosomes have this effect shows that the normal X's play a role in the normal functioning of the nerves and muscles.

A considerable number of other sex-linked characters are known in man. But man is a very unfavourable organism for the study of inheritance. Yet even the little that is known of sex-linked inheritance in man shows that the X-chromosome plays a role in many diverse bodily functions.

For a more complete idea of its role, some organism that can be bred experimentally must be examined. For this purpose the fruit-fly, Drosophila melanogaster, is the best organism to select, since it has been studied more extensively than any other.

In the fruit-fly we find peculiarities of the following types that are dependent on alterations in particular X-chromosomes:

Many different eye colours: The normal eye programmes in this organism is a certain shade of red. Defects in different X-chromosomes result in producing, in the individuals that bear these defective chromosomes, many different shades of red, varying from deep red to a very light red, and thence to 'buff', 'ivory, and 'white'. More than a dozen different types of eye programmes are known to result from modifications of the X-chromosome. It is clear that the normal X-chromosome plays an important role in producing the normal eye programmes.

Structural peculiarities of the eye, such as 'bar-eye', 'facet eye', 'furrowed eye', and the like, are known to result from modifications of the normal X-chromosome.

Wing modifications: Many different conditions of the wings are known that depend upon diversities among X-chromosomes borne by different individuals. These affect all sorts of features of the wings: size, form, venation, function. The normal X-chromosome obviously plays an important role in the full and normal development of the wings.

Body colours and markings: Modifications of the X-chromosomes produce the body colours yellow, sable, tan, chrome, lemon, green, and the like, in place of the normal grey. Other changes in X alter the distribution of pigment on the body, giving the characters dot.

Body structure: A defect in X produces irregularities in the abdomen, known as 'abnormal abdomen'.

Legs: A defect in X-chromosomes causes abnormal development of the legs; some of them are wholly or partly reduplicated, so that the total number of legs is increased.

Bristles: Modifications in X-chromosomes result in various different changes in the bristles that are scattered over the body.

Many physiological conditions and functions likewise depend upon the X-chromosomes, since they are changed when these chromosomes are modified. Among these are the following:

Positive reaction to light: Drosophila with normal X-chromosomes fly toward a source of light. Those having defective X-chromosomes of the kind that produce a tan-coloured bodydo not fly toward a source of light.

Weakness and short life: Most of the, defective X-chromosomes that cause structural or other changes in the body (different eye colours, wing forms, and the like) produce likewise weakness and short life. The individuals bearing them are less resistant to bad conditions, and live for a shorter time than the individuals that bear the normal X-chromosomes.

Life and death: Some X-chromosomes have defects that are so severe that the individuals bearing them will not live and develop, unless there is present also a normal X-chromosome. Such defects are known as lethals. The presence of such lethal defects in the X-chromosome has the result that males bearing them die, since they have but one X; while females usually live, since they have an additional X that is often normal.

If a mother that has one lethal X and one normal one is mated to a father that has a normal X, the result is that half of the sons which receive the mother's defective X fail to develop, while the other half of the sons,receiving the mother's normal X, live and develop.

All the daughters live, since they all receive a normal X-chromosome from the father. The consequence is that in such families there are twice as many daughters as there are sons. In families of 100 to 200, as occur in Drosophila, this is very striking. X-chromosomes are known to play in other organisms roles similar to those mentioned above for Drosophila, although in no other organism has the matter been so fully studied.

We may summarize what has been brought out above in the following statements:

- Many defects that appear in individuals are the result of defects in the X-chromosomes that they bear.
- Some of these defects are dominant; they are manifested in all individuals in which the defective X is present.
- But most such defects are recessive; they are manifested only in individuals in which the defective X is the only kind present.
- When both a normal X and a defective X are present, in most cases the normal X performs the required functions, so that the individual is not defective.
- Since the female has two X-chromosomes, while the

male has but one, such defects are more frequently manifested in the males. Usually one of the X's present in the female is without the defect, so that she is not defective. But when the male carries a defective X-chromosome, the defect is manifested.

- It is thus advantageous, so far as the occurrence of defects is concerned, to have two X-chromosomes rather than one.
- This gives the female a considerable advantage over the male in these respects. There are many different types of defects due to defective X-chromosomes; almost all of them are more common in males than in females. Some of these defects seriously injure the health, or even result in death, if the defective X is the only kind that is present. It is well known that, in general, the death rate is higher in males than in females; this is probably due to the fact that males have but one X-chromosome.
- Among the different individuals of a species are scattered a great number of different types of X-chromosomes having different effects on development. Many of the diverse X-chromosomes are distinctly defective, causing personal defects in the individuals that bear them. Others are not defective.
- Defects or modifications in X-chromosomes affect in different cases all parts and functions of the organism.

THE ROLE OF THE X-CHROMOSOMES IN DEVELOPMENT

What do the facts brought out in the preceding sections show as to the physiological functions of the X-chromosomes; as to their role in development?

For every kind of effect resulting from defects or modificacations in the X-chromosomes, there is a corresponding (opposed or diverse) action of normal or unmodified X-chromosomes. Since defective X-chromosomes of a certain type produce in Drosophila short, ill-developed wings, it follows that the normal X-chromosomes act on the development in such a way as to give long, well-developed wings.For if we substitute the normal X for the defective one, this causes the normal well-developed wings to be produced.

Similar reasoning applies to all the defective conditions that result from defective X's; the normal X's act in such a way as to produce the corresponding normal conditions. It follows therefore that the normal X-chromosomes play a role in the production of eye programmes and structure, body programmes and structure, in the development of the legs, the wings, the bristles, in producing normal health, resistance and vigour, and in various physiological processes. We know, further, that they play a most important part in determining sex, with all that this includes of structural and physiological influence.

Clearly therefore the X-chromosomes play a most important part in development. Their action is not limited to any single part of the body, nor to any single class of functions. They enter into the processes of development in such a way as to influence all parts of the body, all functions of the body. They begin to influence development very early, as was seen in considering the development of sex. And they play important roles in such relatively late processes as the production of eye colours.

It can hardly be doubted that they enter into the developmental activities from practically the beginning, influencing all the later processes that occur.

AUTOSOMES AND T-CHROMOSOMES

The X-chromosomes have been dealt with because of the great advantages they offer for experimental study. It was seen that they affect development in many ways, and that in a given species there are many different types of X-chromosomes in the different individuals, causing them to show different characteristics, structural and physiological. Is the X-chromosome typical in these relations? Shall we find similar effects in the other chromosomes? In addition to the X-chromosomes, there are the autosomes, which are present as a rule as several or many pairs. In man there are 23 pairs of autosomes: in Drosophila there are three pairs. There is, further, in many species the single Y-chromosome. The autosomes and the Y-chromosome are distributed from parent to offspring in characteristic ways, differing from the distribution of X-chromosomes.

If they affect characteristics, these differences should appear in a different system of inheritance. We shall examine first the autosomes.

THE AUTOSOMES: TYPICAL MENDELIAN INHERITANCE

We have already seen that the autosomes play a role in development, in the fact that they help to determine the sex of the developing individual. In Drosophila, if the number of sets of autosomes, in relation to the number of X's, is changed, this changes the sex.

Do autosomes also affect other characteristics? Do the different pairs of autosomes have different functions? Are there diverse types of autosomes, with different effects, in different individuals of a species, as is the case in relation to the X-chromosomes? Are there dominant and recessive effects of autosomes, as there are of X?

If dominant and recessive characteristics depend on autosomes, then these characteristics must follow the distribution of the autosomes from parents to offspring, as the characteristics dependent on X follow its distribution. We must then examine the method of distribution of the autosomes whether there are characteristics that show this method.

METHOD OF DISTRIBUTION OF THE AUTOSOMES

The distribution of the autosomes from parents to offspring is much simpler than that of the X-chromosomes. The main facts are as follows:

- The autosomes are in pairs in all individuals. There

are no individuals with unpaired autosomes, like the male of some organisms with relation to X.

- Thus the two sexes are alike as to their autosomes.
- In forming germ cells, one member of each pair of autosomes goes to each germ cell. The sperms and the ova are alike with respect to the autosomes.

  Thus the fundamental rule with relation to the autosomes is that each germ cell carries one member of each pair of autosomes.
- By union of two germ cells the pairs of autosomes are restored.
- Thus each one of the offspring carries in each pair of its autosomes one autosome from each parent.

Now consider a single pair of autosomes, such as the pair marked AA in. Suppose that in different individuals the members of this pair of autosomes produce different effects, as we have seen to be the case with the X-chromosomes of different individuals. And suppose that, as in the case of X, one of these effects is dominant, the other recessive. How would such characteristics be inherited?

This question must be examined with care. For this purpose we may take as an example the different colours found in different varieties of peas, used in the classic experiments of Mendel. The yellow seed programmes of a certain variety is dominant, and results from the constitution of the autosomes of a certain pair. The green seed programmes of another variety is recessive, owing to a different constitution of the autosomes of that same pair.

Call the two autosomes of that pair in the dominant yellow race by the capital letters *AA*, while the two autosomes that give rise to the recessive green programmes in the other race may be called *aa*. The dominant parent may be designated *D*, the recessive parent *R*, When these two parents produce germ cells, each germ cell receives, of course, one chromosome for the pair. The germ cells from the dominant parent *D* have the autosome *A*, those from the recessive parent *R* have the autosome *a*. The germ cells *A*, from parent *D*, unite each with a germ cell a from parent *R* giving offspring (zygotes) containing both *A* and *a*. These first generation offspring are commonly designated *Fi* (first filial generation).

And since the characteristic carried by the autosome *A* is dominant, all these *Fi* offspring manifest the dominant characteristic *D* (yellow, in case yellow-seeded peas are crossed with green-seeded peas).

Next observe the result of mating together two of these individuals *Aa*, of the *Fi* generation, giving the *Fa* (second filial generation).

Each parent *Aa* gives, according to the general rule, two kinds of germ cells, one kind carrying the dominant autosome *A*, the other the recessive autosome *a*. Each kind of germ cell from one parent unites with each kind from the other. That is, half the *A* germ cells from one parent unite with *A* germ cells from the other parent, half with a germ cells from the other parent; and the same is true for the a germ cells from the first parent. This gives in equal proportions *Fa* offspring (zygotes) as follows:

$$AA + Aa + a\ A + aa$$

But as the two combinations *Aa* and *aA* contain the same chromosomes, the proportions may be written *AA* + 2*Aa-f-aa*

The individuals of the constitution AA and Aa contain the dominant chromosome *A*; therefore they manifest the dominant characteristic, *D*. The individuals *aa*, having only the recessive chromosome *a*, manifest the recessive characteristic, *R*. Thus in this F^2 generation, so far as manifested characters are concerned, the proportion of offspring are 3 Dominant to *i* Recessive

Thus if dominant and recessive characteristics depend on the two members of a pair of autosomes, when a dominant individual *AA* is mated with a recessive individual *aa* the general result is bound to be that the immediate offspring (*Fi*) all manifest the dominant characteristic, while in the *Fa* generation (produced by mating two individuals from *Fi*) there will be three times as many dominant individuals as there are recessives.

Now, this is the result that was discovered by Mendel in 1866; it formed the foundation for the working out of Mendelian heredity. The proportion 30 to *iR* in the *Fa* generation is commonly known as 'the Mendelian ratio'; it is one of the most characteristic proportions in this type of inheritance.

Mendel discovered that there are many different characteristics that yield this 3 to *i* ratio. If yellow and green peas are crossed, in the *Fa* generation there are 3 yellows to *i* green; if round peas and wrinkled peas are crossed, in F^2 there are 3 round to *i* wrinkled; and so for many other characteristics in the pea plants with which Mendel worked.

And since the time of Mendel these same ratios have been found for thousands of different characteristics, in great numbers of different organisms, from the pea plant and *D* rosophila to man. The fact that inheritance gives these ratios is commonly spoken of as Mendel's law, and the type of inheritance which it exemplifies is spoken of as Mendelian inheritance.

The type that gives in the *Fa* generation 3*D* to *iR* is the commonest method of inheritance, much commoner than the sex-linked method, in which the characteristics depend on differences in the X-chromosomes. This is obviously because autosomes are more numerous than X-chromosomes.

We find then that inheritance according to Mendel's law takes place exactly as it would if the characteristics concerned depended on the two members of a pair of autosomes. As will be shown later, it can be proved experimentally that such characteristics do indeed depend upon autosomes.

Thus the rules of Mendelism are the rules of distribution of the two members of a pair of autosomes, one having a recessive effect, the other a dominant effect. Mendelian heredity is heredity of characteristics that depend on differences in autosomes.

Mendel did not know of the relation of the characteristics to the chromosomes. He discovered, however, certain other important facts about this type of heredity, facts which have become clearly explicable since it has been learned that this type depends on differences in autosomes. These facts must be examined.

For this purpose it will be convenient first to define certain terms which are universally employed in dealing with inheritance.

By the union of two germ cells or gametes, as *A* and *a*, there is produced a zygote. If the two germ cells are alike with respect to the chromosomes or characteristics that they carry, the zygote so produced is spoken of as a homozygote; thus *AA* and *aa* are homozygotes. If the two germ cells that unite to form a zygote are unlike, as *A* and *a*, the zygote so produced is called a heterozygote, as *Aa*.

The two corresponding characters, one dominant, the other recessive, that are produced by the two differing members of a pair of chromosomes are spoken of as alternative characters, or alleles, or allelomorphs. Any given individual manifests one or the other of the two alternatives, not both. For the alternative characters we shall employ henceforth the term alleles.

The characteristic manifested by a zygote is its phaenotype while its actual constitution is its genotype. Thus, two zygotes *AA* and *Aa* both manifest the dominant character; the dominant character is the phaenotype, or the individuals are said to be phaenotypically dominant. But they are diverse in genotype; one is *AA*, the other *Aa*. In the case of a homo-zygote *AA* or *aa*, the phaenotype and genotype may be held to coincide.

Mendel crossed yellow peas with green peas, and found, as before stated, that in the *Fi* generation all the peas were yellow, so that yellow is the dominant allelomorph. In the F^2 there were three times as many yellows as greens. The actual number of seeds in the F^2 generation was in his experiments 8023, of which 6022 were yellow while 2001 were green, so that the ratio of dominants to recessives was, accurately expressed, 3.01 to *i*. Mendel found that if he bred further the individuals of the F^2 generation, some of these gave different results from others. In the pea plant it is possible to allow each plant to fertilize itself, so that both kinds of germ cells come from the same parent. When this was done with the individuals of the F^2 generation, the following results were observed: The green plants (recessive) gave only recessive green offspring (they 'bred true').

Of the yellow plants (dominant), one-third gave only yellow offspring (they 'bred true'), while two-thirds of them gave both yellow and green offspring, in the proportion of 3 yellows to *i* green.

The reason for this is clear when we consider the chromosomes on which the characters depend. If *A* represents the chromosome (autosome) that gives the dominant yellow programmes, while a represents the chromosome that gives the recessive green, the original parents were *AA* and *aa*, the *Fi* generation were all *Aa*, while the *Fa* generation showed the proportions *AA-f* 2 *Aa* + *aa*. The recessive individuals *aa* are homozygotes; their two chromosomes are alike, being *aa*. When germ cells are produced by these homozygotic individuals, all these germ cells must contain the chromosome *a*, and when two of these unite in self-fertilization, of course they produce in all cases again individuals *aa*. These are, of course, recessives (green), like their parents; that is, the green parents aa 'breed true'.

Similarly the dominant yellows, *AA* (constituting one-third of all the dominants), are homozygotes. Their germ cells all contain the chromosome *A*, and the offspring produced by union of such germ cells, in self-fertilization, are all *AA*, and so dominant yellow like the parents. One-third of all the dominants thus breed true.

But the individuals *Aa* are heterozygotic; that is, they have two kinds of chromosomes, *A* and *a*, in the pair. Thus they produce two kinds of germ cells, *A* and *a*, in equal numbers. And when these germ cells unite, they necessarily give, as we saw on an earlier page, offspring of three kinds, in the proportions *AA* + 2*Aa*4-*aa* Of these, *AA* and 2*Aa* will be dominants (yellow), while *aa* will be recessive (green). Thus the homozygotes *aa* and *AA* breed true, while the heterozygotes, *Aa*, do not.

As Mendel discovered, if the heterozygotes *Aa* are mated

with the recessive homozygotes *aa*, they give two kinds of offspring, *Aa* and *aa*; half (*Aa*) being dominant,the other half (*aa*) being recessive. If the heterozygotes *Aa* are mated with the dominant homozygotes *AA*, the offspring *AA* and *Aa* are all dominant, though half (*AA*) are homozygotes, the other half (*Aa*) heterozygotes.

All these results are typical for the descendants of two parents different in a single pair of chromosomes, one pair being dominant, the other recessive. For reference it will be well to make a table of these results, characteristic for Mendelian inheritance.

THE RESULTS OF MENDELIAN INHERITANCE

Parents: *AA* and *aa*; both homozygotic. *Fi*: *Aa*; all alike, dominant heterozygotes.

Fs: *AA* + 2*Aa* + *aa*. Three-fourths of the individuals dominant, onefourth recessive. Half the individuals homozygotes (*AA* and *aa*), the other half heterozygotes (*aAa*).

Each type of homozygote bred by itself produces offspring like itself.

The heterozygotes *Aa*, when bred by themselves, produce again offspring

AA + 2*Aa* + *aa* *Aa* mated with aa yields again *Aa* + *aa* *Aa* mated with *AA* yields again *Aa* + *AA*. All of these results are found to hold true for thousands of different characteristics.

We may therefore summarize the situation in the following general propositions:

- When two different individuals are crossed, in many cases the differences between them follow the same method of distribution as do the two members of a pair of autosomes ýÿone being dominant, the other recessive.
- This distribution is what is known as typical Mendelian heredity.
- The two sexes do not differ in these respects. Each sex has the same proportion of dominants and recessives.

- The rules of Mendelian inheritance are essentially the rules of distribution of the members of pairs of autosomes.

Characteristics that depend on two or more pairs of autosomes.We have thus far followed the distribution of characteristics that depend on differences in a single pair of autosomes. But most organisms have several or many pairs of autosomes: the fruit-fly has three pairs, man has twenty-three.

Do some characteristics depend on differences in one pair of autosomes, others on differences in other pairs? Have the different pairs of autosomes thus different functions in development and inheritance?

We find that the different pairs of autosomes have indeed different functions, and that some characteristics depend on one pair, some on another. This is shown in many cases in which two parents differ in two diverse characteristics. One characteristic may then act as if dependent on one pair of autosomes, the other on another pair. As an example of this, we may take a case studied by Mendel in peas.

One of the original pea plants had seeds that were (i) yellow, and (2) round, or smooth. These, as it turns out, are both dominant characters. The other plants, crossed with those just mentioned, had recessive characters; the seeds were (i) green, and (2) wrinkled.

When these two kinds of pea plants were mated together, it turned out, as we shall see, that yellow and green behave in inheritances as if due to a difference in the two parents in the effects of a certain pair of autosomes; that is, they are alleles. Round and wrinkled behave as if due to a difference between the two parents in the effects of another pair of autosomes.

We may therefore represent the two pairs of autosomes. Representing the dominant characters by capitals, the recessives by lower-case letters, we may in the diagram represent the chromosomes giving origin to the four characters as follows:

A = yellow a = green

B = round *b* = wrinkled

The parents will then be represented as shown in Fig. The germ cells from these parents contain, of course, one member of each pair, as shown in Fig. One parent produces gametes all of the constitution *AB*, the other gametes all of the constitution *ab*. When a gamete *AB* from one parent unites with a gamete *ab* from the other, all the offspring (*Fi*) thus produced have the constitution *AaBa*, as shown at *Fi* in Fig.. Since the dominant autosomes *A* and *B* are both present, all these individuals are dominant for both pairs of characters; that is, they all have seeds that are yellow and round, as Mendel found. But all are heterozygotes-for both of these characters.

Next we mate together two of these *Fi* individuals to give the F^2 generation. In forming gametes the rule is, of course, that each gamete contains one member of each pair of chromosomes. But the two pairs, *Aa* and *Bb*, are distributed to the gametes quite independently. Thus, some get *A* and *B*, some *A* and *b*, some a and *B*, some *a* and *b*.

There are thus from each parent four types of gametes, *AB, Ab, aB*, and *ab*, each type being represented by many gametes. The gametes of each of the four types from one parent unite in approximately equal proportions with gametes of each of the four types from the other parent. That is, gametes of the type *AB* unite in different cases with *AB, Ab, aB* or *ab* from the other parent; and so of each of the other gametes of the first parent. The result is to give a set of 16 combinations of gametes (some of which are alike, however, as will be seen). It is not necessary to carry further in our diagram the outlines of the chromosomes.

INHERITANCE DEPENDING ON TWO PAIRS OF AUTOSOMES

AB AB AB AB

AB(i) *Ab*(6) *aB*(7) *ab*(s)

Ab Ab Ab Ab

AB(6) *Ab*(2) *aB*(5) *ab*(8)

aB aB aB aB

$AB(_7)$ $Ab(_5)$ $aB(_3)$ $ab(_9)$

ab ab ab ab

AB(5) *Ab*(8) *aB*(9) *ab*(4)

But, as will be observed, not all these 16 are diverse; thus *ABAb* is the same in constitution or genotype as *AbAB*. If we number the different combinations, and give the same number to those that are alike, as is done (in parentheses) we find that there are nine different combinations or genotypes. That is, two parents that are both *ABab* produce offspring of nine different constitutions or genotypes.

When the parents differed in but one pair of characters there were but 3 different combinations. When there are two pairs the number is 3 > or 9⁻.

These nine different combinations will manifest the dominant characters *A* or *B*, in all cases where an *A* or *B* is present whether present once, as *A*, or twice, as *AA*); they will manifest the recessive characters *a* or *b* in all cases in which *a* or *b* is present twice, no corresponding capital letter being present. Collecting together these different types, we find the different characteristics to be manifested in the following proportions of the zygotes:

9*AB* + 3*Ab* + *saB* + 1*ab*

That is, 9 out of 16 are dominant for both pairs of characters, 3 are dominant for *A*, recessive for *b*; 3 dominant for *B*, recessive for *a*, and *i* out of 16 is recessive for both. There are thus 4 different phaenotypes. In Mendel's experiment with peas, in which the two pairs of characters were (1) yellow and green; (2) round and wrinkled; in *Fi* the proportions were 9 that were yellow and round, (3) that were yellow and wrinkled, (4) that were green and round, and *i* that was green and wrinkled. These proportions hold generally when there are two pairs of characters connected with different pairs of autosomes. It may be observed that the proportions of the individuals of different constitution or genotype in F^2, descended from a cross between parents that differ in respect to two pairs of autosomal characters (so that the parents are *AABB* and *aabb*), can be found directly from the proportions for a single pair of characters.

In *Fa*, for the single pairs the proportions are *AA* + 2*Aa*-

haA and *BB* + 2*Bb* + *bb* respectively. If now we multiply together algebraically these two polynomials, we obtain the relative proportions for the different combinations produced when the two pairs are together. These proportions are the following:

$$AABB + 2AaBB + aaBB + 2AABb + 4AaBb +$$
$$(1)\ (7)\ (3)\ (6)\ (5)$$
$$2aaBb + AAbb + 2Aabb + aabb$$
$$(9)\ (2)\ (8)\ (4),$$

Certain other relations among the nine different types are important, since they hold for all cases of two pairs of characters that are dependent upon different pairs of autosomes. These are the following:

- There are four combinations (Nos. 1, 2, 3, 4), constituting one-fourth of all the zygotes, that are homozygotes for both pairs of genes. The four homozygotes are diverse in their combinations and in the characters they manifest. Any one of these when self-fertilized or mated with another like itself breeds true: that is, it produces offspring that have the same constitution as the parents.
- Four out of the 16, or one-fourth of all the zygotes, are heterozygotes for both pairs of characters. They form the class numbered (5), and constitute the diagonal from upper right to lower left. Any of these when self-fertilized or mated with another like itself gives again the nine different types, in the proportions above set forth.
- The one-half of all the zygotes, are homozygotes for one of the two pairs, heterozygotes for the other. Four are homozygotic for *A* and heterozygotic for *B*; the other four are heterozygotic for *A* and homozygotic for *B*. In these 8 there are but 4 diverse combinations or genotypes.

If any one of these is self-fertilized, it yields 3 diverse types with respect to the pair for which it is heterozygotic. Thus, the class (6), having the constitution AABb, yields when self-fertilized three combinations in the following proportions:

$AABB + 2AABb + AAbb$

The above relations hold for all cases in which two pairs of characteristics are dependent on differences in two different pairs of autosomes, whatever the method of designating the characters or chromosomes.

In a similar way, two parents may differ in three characters that are dependent on three different pairs of autosomes. Such parents could be represented as *AABBCC* and *aabbcc*.

Their progeny (*Fi* generation) would all have the constitution *AaBbCc*. Such *Fi* individuals of course produce a variety of germ cells, in accordance with the principles that each germ cell receives one member of each pair of autosomes, and that the different pairs of autosomes are distributed independently. Each parent *AaBbCc* thus yields the following 8 types of germ cells:

ABC, ABc, AbC, aBC, Abe, aBc, abC, abc.

When each of the 8 kinds from one parent unites with each of the 8 kinds from the other, there are produced, of course, 64 groups. If we classify the 64 individuals by the characters which they manifest, we find that there are 8 different phaenotypes manifesting respectively the following characteristics:

ABC, ABc, AbC, aBC, Abc, aBc, abC, and *abc*

That is, some are dominant for all three characters; some for a particular two and recessive for the other; some are dominant for a particular one and recessive for the other two; and some are recessive for all three. The different phaenotypes are present in different proportions; the proportions when a large number are examined are as follows:

2*ABC* + *gABc* + *gAbC* + *gaBC* + $*Abc* + *saBc* + 3*abC* + *iabc*

The total number of different combinations (genotypes) that occur in such a case is. These combinations can be obtained in their correct proportions by multiplying together algebraically the expressions: Similar but still more complex series are obtained if parents are interbred that differ in four or more characters depending on four or more different pairs of autosomes. If the number of pairs of autosomes is called ↔, the following relations hold:

If parents differ in n autosomes:

- The number of diverse types of gametes produced by the individuals of the *Fi* generation is 2^n.
- The number of different combinations (genotypes) in the zygotes of F^2, formed by mating the 2^n gametes from *Fi*, is s^w.
- The number of different phaenotypes produced in *Fa* is 2^n.

These relations yield very great numbers of diverse combinations, in cases where the number of pairs of autosomes is large. Thus in man, with 23 pairs of autosomes, if all the autosomes differed in the two parents, the number of types of germ cells produced by the *Fi* individuals would be 2, or many millions, the number of possible diverse genotypes in F^2 would be 3, or many billions, and the number of possible different phaenotypes would be 2. (As will appear later, the number of different possible combinations is greatly increased by the exchange of parts by the members of a pair of chromosomes.)

Thus typical Mendelian inheritance, dependent on differences between different pairs of autosomes, can be treated to a certain extent mathematically. The individuals in F^2 and later generations occur in proportions represented by simple mathematical ratios.

It will be worth while to formulate the main facts that we have brought out as to autosomal inheritance:

- Some single pairs of characters, dominant and recessive, follow in inheritance the method of distribution of a single pair of autosomes.
- Some pairs of inherited characters follow the rules of distribution of two or more different pairs of autosomes, in the following particulars:
  - One member of each pair goes into each germ cell.
  - The members of the different pairs are distributed to the germ cells independently, so that all possible combinations containing one

member of each pair are formed with equal frequency.

  - Each type of gamete from one parent unites with each type from the other parent, the different matings being equally frequent.

- These methods of distribution result, for the characters manifested in the Fs generation, in certain typical numerical proportions or ratios, known as Mendelian ratios.
- This method of inheritance is known as Mendelism, or Mendelian inheritance. The rules of Mendelism are the result of the rules of distribution of pairs of autosomes.
- Most characteristics of organisms are inherited in this way; many more than are inherited in the sex-linked manner. This is evidently because there is but a single pair of X-chromosomes, while there are several or many pairs of autosomes.
- Thus it is clear that, among the different individuals of a species, there are many different types of autosomes (as there are different types of X-chromosomes), producing many kinds of characteristics.

THE T-CHROMOSOME

There remains another chromosome, not included in the autosomes: namely, the Y-chromosome. Has this functions similar to those of X and of the autosomes? Does it too play a role in heredity?

As we shall see, the Y-chromosome differs in many ways from the other chromosomes. Let us first summarize the main facts as to the occurrence and distribution of Y-chromosomes.

- Some organisms have no Y-chromosomes. Among those are many insects the dog, the cat, the horse.
- In some organisms it is small and appears to be degenerate. This is the case in many insects; also in man.

- In some organisms the Y-chromosome is large, but of a different form from its mate X. Such is the case in Drosophila
- In some organisms Y is not distinguishable from X inform or size, but experiment shows that it differs from X in function.
- The Y-chromosome descends from father to son only, exclusively in the male line. Normally it never occurs in females.
- Hence any characters that depend on the presence of a particular kind of Y will *(a)* be found in males only, and will *(b)* never be inherited *through* females (since Y does not pass through females).

THE T-CHROMOSOME IN DROSOPHILA

The functions of the Y- chromosome have been more thoroughly studied in Drosophila than in any other organism. Most of the facts as to this have already been brought out, in connection with our account of the functions of the X-chromosome. We may summarize as follows the known facts as to the functions of the Y-chromosome in Drosophila:

- The Y-chromosome, although it normally occurs in males, is not required for the production of male individuals. Through non-disjunction, male individuals are produced that contain no Y.
- But the presence of Y is necessary for the fertility of males.
- Y may exist in females, as a result of accidental nondisjunction; it does not affect their sex.
- During many years of intensive study no characteristics were found that follow the Y-chromosome.
- But recently it has been shown that the presence of Y prevents the appearance of a certain recessive character ('bobbed') that is due to a modified X-chromosome. This modified X results in the

production of shortened bristles on the body of the fly. But this peculiarity does not occur if a Y is present, so that Y has the dominant normal allele for 'bobbed'. Indications of certain other ill-defined functions of the Y-chromosome have of late come to view.

- No diverse modifications or defects such as are known in great number for the X-chromosome are known in the Y-chromosome.
- Thus in Drosophila the Y-chromosome appears to have little function; it seems nearly but not quite inert.

Active T-chromosomes in certain Organisms; Relation to Male Secondary Sex Characters. In certain organisms the Y-chromosome is more active and varied in its effects. In such cases it shows in different individuals diverse modifications, comparable to the diverse types of the X-chromosome in Drosophila, and these produce diverse inherited characteristics that appear in males only.

This matter is closely bound up with the question of what produces the diversity of secondary sex characters. In many animals there are special characteristics which, like the Y-chromosome, are found only in males.

These are the male secondary sex characters: horns, mane, beard, greater size, diversities in form, and the like. Since these appear only in the male, the question arises as to whether they may be due to the presence of the Y-chromosome. With relation to this the following is the situation:

- In most cases, the presence of male secondary sex characters does not depend on the presence of the Y-chromosomes. This is shown by the following:
  - There are animals which have no Y-chromosome, yet have male secondary sex characters: for example, the dog, the cat, the horse; also various insects.
  - In Drosophila, males that contain no Y (as a

result of non-disjunction) have the typical male secondary sex characters.

— In some breeds of sheep the male has horns, while the female has none. It could be suggested that this is due to the fact that the male carries the Y-chromosome, while the female does not. But in other breeds, as Dorsets, both sexes have horns, although only the male has the T-chromosome. And in still other breeds, as Suffolks, neither sex has horns, although here as elsewhere the males have the Y-chromosome.

If a female of the horned breed (Dorset) is crossed with a male of the hornless breed (Suffolk), in the Fi descendants the males have horns, although they receive their Y-chromosome from the hornless breed. They have inherited the horns through the mother.

Thus it is clear that in these cases the presence of horns is not due to the presence of the Y-chromosome. Extensive work on the inheritance of horns and horn- lessness in sheep, goats and cattle consistently gives results which show that the presence of horns is not due to the presence of the Y-chromosome, or of any special type of Y-chromosome.

In all these cases (a), (b) and (c) the presence of distinctive male secondary sex characters depends in some way on the fact that the male carries but one X-chromosome, and consequently produces the male hormone. Their absence from the female results from the fact that she carries two X-chromosomes, and consequently produces the female hormone.

- But in certain fish, some of the male secondary sex characters apparently do depend on the presence of a Y of a particular type, for they occur only where that type of Y is present, and they can therefore not be inherited through the mother. That is, the Y-chromosomes of certain individuals or breeds are so modified as to produce certain definite characteristics in the individuals that contain them, while individuals that contain another type of Y do not show these characteristics.

An example of this is found in the small fish Lebistes. In a certain race, the males have a black spot at the base of the dorsal fin, while the females are without the spot. In another race of the same species neither the males nor the females have the black spot. Thus the females of the two races are alike in lacking the black spot, while the males differ. Is this difference the result of differences in the Y-chromosomes of the two kinds of males?

This may be tested by crossing members of the two races. In the spotted race, represent the X and Y-chromosome by capital letters; while in the non-spotted race, represent them by the small letters *x* and *y*.

A female of the spotted race, XX, is mated with a male of the non-spotted race, *xy*, thus:

Females Males
Parents XX *xy*
Non-spotted
Fi Xx Xy
Non-spotted

THE Y-CHROMOSOMES

This produces in Fi females Xx, and males Xy. Neither contains the Y from the spotted race, and neither is spotted. The spot is not inherited through the females of the spotted race.

Next mate the female of the non-spotted race, xx, with a male from the spotted race, XY, as follows:

Females Males
Parents xx XY
Spotted
Fi xX xY
Spotted

This yields in Fi females Xx, and males xY. These males carry the Y from the spotted race, and they are spotted. Thus far the spot goes with the Y from the spotted race.

This Y may be traced further, for example, by breeding together the non-spotted females xx with the Fi males xY, from the preceding cross, thus:

Females Males

Parents xx xY

Spotted

F2 xx xY

Spotted

This gives in the Fs females xx, and males xY. These males are again spotted; they carry the Y from the spotted race. Wherever this Y from the spotted race is found, the black spot appears; wherever the Y is from the non-spotted race it does not appear. It appears clear therefore that whether the black spot appears in the male depends upon the type of Y-chromosome that it carries. There are two types of Y-chromosome in the species, one producing the black spot, the other not producing it.

It has been shown by Winge that there are many other programmes markings in this fish that depend for their occurrence on the presence of a particular type of Y-chromosome. They occur only in the males, and only in males that have a certain type of Y. In certain other fish also the Y-chromosomes have been shown to be diverse in different individuals, giving rise to distinctive characteristics in the male.

In a certain plant, Melandrium, also, there are inherited characteristics that depend upon the presence of a particular type of Y-chromosome.

There are some indications also that in man the nature of the Y-chromosome affects certain characteristics. In a family described by Schofield, the male parent in the first observed generation had webbed toes. This was inherited in the male lines only by all the males for four successive generations. It thus followed exactly the distribution of the Y-chromosome from the original male parent. It seems probable therefore that it is due to a defect or modification in the Y-chromosome of that individual. Many other cases of webbed toes or fingers are known that occur in both sexes, so that they are clearly not due to the Y-chromosome.

On the whole, the situation as to the Y-chromosome and its relation to inheritance may be summarized as follows:

- In most organisms Y is a degenerate chromosome, lacking most of the functions manifested by the other chromosomes.
- In some organisms degeneration has gone so far that the Y-chromosome has completely disappeared.
- In other cases it exists, but is very small.
- But in some species the Y-chromosome has effects on hereditary characters that are similar to those produced by other chromosomes. Different types of inherited characters result in such cases from diversely modified Y-chromosomes.
- Characteristics dependent on the presence of a particular type of Y-chromosome are inherited in the male line exclusively.

THREE MAIN METHODS OF DISTRIBUTION OF INHERITED CHARACTERISTICS

There exist therefore three diverse systems of distributing hereditary characteristics, depending respectively on the X-chromosomes, the autosomes, and the Y-chromosomes. In each of these the rules of inheritance are given by the special method of distribution of the three types of chromosomes, taken in connection with the facts of dominance and recessiveness.

Inheritance which follows the distribution of the X-chromosomes is commonly called sex-linked inheritance. That following the distribution of autosomes is known as autosomal or typical Mendelian inheritance. Less important than those just named is Y-chromosome inheritance. All these methods are included under the term Mendelian inheritance, employed in a broad sense.

In some organisms still other methods of inheritance exist.

In higher animals and plants most inheritance occurs in one of the three methods named in the preceding paragraph.

CHARACTERISTICS AFFECTED BOTH BY GENES AND BY ENVIRONMENT

Many of the same features that can be altered by changing the genes can likewise be altered by appropriate changes in the environment. Characteristics do not fall into two classes, one exclusively hereditary (or dependent on genes), the other exclusively environmental; but any characteristic is affected both by the materials of which the organism is composed, and by the action of the conditions on these materials.

Yet, as we shall see, in practice some characteristics are more readily altered by environmental conditions than are others. Certain characteristics owe most of their peculiarities to diversities among genes. Others are readily affected both by genes and by environment. Still others depend mainly on environmental conditions.

Particularly illuminating are the cases in which characteristics depend in marked degree both on genes and on environmental conditions, so that they are altered by changes in either. A number of well-known cases of this kind will be presented, selecting those that are from a technical point of view fully known.

In that animal whose genetics is best known, the fruit-fly Drosophila, many of the characteristics that have been studied are defects or abnormalities which are sharply dependent on the presence of particular genes. Such a one is abnormal abdomen. This appears in the fact that the abdomen is ill formed, the segments not being regular nor sharply marked off. This abnormality is found to be due to a defect in a certain gene of the X-chromosome, so that it shows sex-linked inheritance, the abnormality being dominant.

Unless that particular defective gene is present the abdomen is normal. But it is dependent too on the environment. Individuals with the abnormal gene, if developed in a moist atmosphere, have the abnormality. But if those same individuals are developed in a dry atmosphere, they have normal abdomens. To produce the abnormal abdomen, we must have first the particular kind of defective gene that yields it. But second, even if that is present, we must have, too, the moist atmosphere for development; without it the abnormality is not produced, whatever the gene.

Another character in the fruit-fly that acts in a similar way is what is called reduplicated legs. If a certain defective gene is present in the X-chromosome, the animals show a tendency to produce double legs, or even triple or multiple legs, in place of single ones. But this does not happen unless the young develop at a low temperature. To produce reduplicated legs, then, there must be present a certain type of gene, and also a certain external condition. Unless both are present, the peculiarity is not produced.

It is clear from these cases that even though an individual inherits the type of gene necessary to produce a certain defect, it may not be inevitable that he should have that defect. A special environment may correct what inheritance leaves imperfect. These considerations apply to many things in man as well as in other organisms.

Another example of the fact that what is produced in development depends both on the genes and on the environment is seen in the production of giants in the fruit-fly. A modified gene, located in the X-chromosome, near its left end, causes the animals (if they have no other type of X-chromo- somes than this) to become giants; they grow to nearly twice the size of the ordinary fly.

But this increase in size takes place only if the animals having the modified gene are well fed during a certain period of their early lives. If not well fed at this particular time in life, they grow no larger than the usual flies. The giant size requires to produce it a particular type of environment acting on a particular type of gene. If either condition is not fulfilled, giants are not produced.

Again, in plants, the difference between green plants, containing chlorophyll, and white ones, is in some cases due to a difference in genes. Plants with a certain kind of genes will remain white even though grown in the light. In other cases, the difference between white and green plants is not the result of a difference in genes, but of a difference in environment.

Plants that live in the dark remain white, even though they contain genes that can produce chlorophyll, while similar plants grown in the light are green. That is, to produce chlorophyll, a certain type of genes is necessary, and also a certain type of environment.

Again, there are red and white varieties of primroses; the difference in programmes is inherited; it is due to gene diversities. But a certain variety produces red flowers when grown in a cool place, white flowers when grown in a warm, moist region, as in a greenhouse. The same diversity in programmes is now due to an environmental difference.

Complex interrelations of genetic and environmental action have been described by Emerson with relation to diverse colours in maize plants. In different types, the leaves, the husks, the flowers, and other parts, differ in programmes. Emerson describes and Figs six main types, under each of which are subtypes. These are the following:

- Purple, with its subtype, weak purple.
- Sun-red, with a subtype, weak sun-red.
- Dilute purple.
- Dilute sun-red.
- Brown.
- Green, with a number of subtypes.

The interrelations of genetic and environmental factors in these colours may be summarized as follows:

- Each programmes and each subtype is constant and hereditary, when the different types are bred under like conditions.
- Each of the colours depends on many different genes, like the eye programmes in Drosophila; they may be altered by changes in any of these genes. The different colours result from different combinations of the genes.

- When the different types are crossed they give typical Mendelian inheritance: in some cases with simple ratios, in others with complex ratios.

  Thus far the conditions are those for typical hereditary characters; the differences are the result of gene differences, and they follow the distribution of the different types of genes in inheritance.

  But, also, the colours depend on the conditions to which the plants are subjected while devéloping. Change of certain conditions changes the colours, as does alteration of the genes. Some of the relations are the following:

- The 'sun-red' type requires exposure to the sun for production of the red programmes. Parts protected from the sun are green. In the 'green types, on the other hand, subjection to the sun does not cause the production of red programmes. Thus the difference between red and green plants is due in some cases to diversity of genes, as when 'sun-red and 'green types are both grown in the sun. In other cases the difference between red and green plants is due to environmental difference, as when 'sun-red plants grown in the sun are compared with those of the same type grown in the shade.

  The plants of the 'purple' type do not depend on exposure to the sun for their programmes; they are purplish in programmes whether grown in sun or shade.

  The degree of programmes produced by exposure to the sun depends on what genes are present. In the typical 'sun-red' variety, exposure to the sun causes a red programmes widely distributed on the plant. In the 'dilute sun-red', exposure to the sun under similar conditions causes only a little red, at the bottom and tips of the leaves, while in the 'green type no red appears.

The nature of the soil affects the programmes. 'Dilute sun-red' grown in poor soil and subjected to the sun has almost all parts red; in good soil, with the same relation to the sun, it has very little red. The 'green and 'brown' types are not influenced in their programmes by the nature of the soil.

- Certain particular chemicals in the soil affect the programmes. Presence of much nitrogenous matter tends to suppress the red; if little is present the red programmes is marked.
- Storage of carbohydrates in the leaves of the programmes-producing varieties causes increase of the red programmes. The amount of carbohydrate in the leaves may be increased by experimental procedures, as by removing the growing ears, or by bending a leaf so as to break the conducting vessels; the leaves thereupon become red in the parts with increased carbohydrate. But this effect is not produced in the 'green' types; they do not produce the red programmes even in presence of stored carbohydrates.

It appears clear, therefore, that in maize different genes cause the different varieties to have different metabolic processes, different chemical reactions, some of which cause the production of colours, while others do not. Also, diverse environmental conditions likewise cause changes in these metabolic processes, inducing programmes production in some cases.Whether programmes occurs, and its type and amount, depend on interaction of genetic and environmental conditions.

Another instructive case of the interaction of genes with environmental conditions is furnished by the conditions determining the number of facets in the compound eye of Drosophila. In the normal eye there are present about 800 facets. In certain of the flies a gene at locus 57 in the X-chromosome is so modified as to cause a great reduction in the number of fully developed facets.

In this condition, known as Bar-eye, there are only about 80 fully developed facets: these form a 'bar' across the eye. This Bar-eye condition is dominant; the number of facets is reduced even though one of the two genes of the pair is normal. We may call the modified gene that causes the reduction in the number of facets the Bar gene.

The number of fully formed facets, it is found, depends on whether there is present but one or two of the Bar genes, and also on whether there is present in addition a normal gene. In the males (having of course but one X), the presence of the Bar gene reduces the number of normal facets to about 90. In the female, if both the X's contain a Bar gene, the number of facets is about 65. If the female is heterozygotic, so that but one of her X's contains the Bar gene, while the other is normal, the mean number of facets is about 358. Finally, in either male or female, if no Bar gene is present, the number of facets is about 800.

Later it was discovered that this Bar gene is sometimes modified in a different way, so as to cause a less marked reduction in the number of facets. This modification is known as infra-Bar. A female containing but one of these infra-Bar genes has about 716 facets; if she contains two she has about 348.

Again, by an extraordinary process, almost or quite unique in known genetics, in some cases two of the Bar genes get united by crossing-over into one X-chromosome. This condition is known as double Bar. When both of the chromosomes carry this double Bar, the number of facets is but about 25. If one carries double Bar, the other single Bar, the number of facets is about 36. Thus the number of facets in the eye depends in many ways on the number and conditions of the genes present. This may be shown in the following diagram, representing the conditions in the two X (or Y) chromosomes of any individual, with the corresponding numbers of eye facets present. In this diagram

+ = the normal, unmodified X-chromosome Y = the Y-chromosome of the male *B* = the Bar gene *iB* = the infra-Bar gene

BB = the double Bar condition (two Bar genes in one chromosome)

Ghr. *i*: + *iB B iB B iB B BB BB BB BB* Chr.s: + + + *iB* Y *BB* + *iB B BB* Facets 800 716 358 348 98 73 65 45 42 36 25

All these various conditions are inherited as unit-difference, sex-linked characters; they are typical hereditary conditions.

But it is found further that the number of facets in most of these conditions depends also on the temperature to which the developing animals are subjected. The numbers given above are those observed by investigators working with fruit-flies cultivated at ordinary, not precisely controlled, temperatures.

If different sets of flies are kept at different temperatures, it is found that high temperatures reduce the number of facets in the individuals that carry the Bar genes, though not in the normal individuals. Krafka found the following mean numbers of facets in the various gene conditions, at different temperatures:

Temperature, Degrees	Bar Females B-B	Bar Males B-Y	Double Bar Females BB-BB	Double Bar Males BB-Y
J5	.214	270	5i	61
20	122	161	33	37
25	81	121	25′	28
300131	40	74	‘5	‘4

Thus the number of eye facets depends both on the genes and on the temperature. The number 61 may be produced either by a double Bar Male (*BB* – Y) at 15 degrees or by a Bar Female (*B* – *B*) at between 25 and 30 degrees. Changing either the genes or the temperature will change the number of facets.

Similar situations occur in vertebrates; one of these may be described briefly. The spotted salamander, Salamandra maculosa, is largely black in programmes, with spots or stripes of yellow. The amount of yellow differs in different cases, so that on the whole some specimens are lighter, some darker. In the lighter individuals the yellow spots tend to run together into rows. The different grades are heritable; certain stocks are regularly darker, others lighter. If darker and lighter individuals are mated, the two conditions show Mendelian inheritance, the darker condition being dominant all the Fi generation are darker, while in Fa there are three dark to one light. The difference between the two thus depends on a difference in one gene.

But it can be shown also that the extent of programmes depends on the environment. If the salamander is kept continuously for a long time (a year or two) on a light background, the yellow spots increase in size: they may run together into stripes or into large yellow areas *B*, so that the animals become much lighter.

If, on the other hand, they are kept for long periods on a dark background the yellow areas decrease in size; the animals become on the whole darker. The extent of the yellow programmes thus depends both on the genes and on the environment; it can be altered by appropriately changing either.

In all these cases the following general principles apply:

- What are really inherited (passed on from parent to offspring) are the genes; that is, certain materials that in certain combinations and under certain conditions give rise to certain definite characteristics.
- With the same original genes, different environmental conditions may induce the production of diverse characteristics.
- Also, with the same environmental conditions, different genes may induce the production of different characteristics.
- The same difference in characteristics that is in one case produced by diversity of genes is in other cases produced by diversity of environment.

There can be little doubt that these general statements apply to many characteristics in man. Some combinations of human genes form an individual that is a much better culture medium for the bacteria of tuberculosis than are others. A person who gets such a combination of genes will develop tuberculosis, if he lives in a region in which the germs of that disease are abundant; while another person, with a different set of genes, will not be subject to tuberculosis, even though living under the same conditions. And the person with the genes that make him susceptible to tuberculosis will not have the disease if he prevents infection by the tubercle bacillus.

In relation to such susceptibility there are doubtless a great number of grades, dependent on just what combination of genes is present. Some are extremely susceptible, others less so but still prone to the disease, and so on, up to individuals whose gene combinations are such as to give them natural immunity to tuberculosis.

What then shall be answered when it is asked whether tuberculosis is hereditary or not? One can only say that some gene combinations predispose to it more than others, so that a hereditary factor is involved. But also there is a necessary environmental factor; it is the interrelation of the two sets of factors that determines whether the individual shall be afflicted with the disease.

Similar considerations apply to any other disease, acute or chronic. Certain combinations of genes are doubtless more readily attacked by plague, by smallpox, by typhoid, by pneumonia, than others, just as certain genetic constitutions yield more readily to extremes of temperature, to exposure to the elements, to unfit food. There is no affair of life in which the gene combination borne by the individual does not play a part. But the environment too plays a part, often an overwhelmingly important part.

The relative role played by genes on the one hand, and by environment on the other, is not the same for all characteristics. Differences in certain features of the organism are mainly due to differences in the genes of which they are composed. In other characteristics, differences are due mainly to environmental diversities. There are all sorts of intermediate conditions between these extremes. We may pass in review a number of different types of characters from this point of view.

Most of the diversities in obvious physical features in most animals are the result of differences in the genes, so that they are mainly matters of inheritance rather than of environment. In Drosophila the forms of the parts of the body, of the legs and wings, the distribution of bristles, the venation of the wings, the colours of body and eyes, are all changeable through the action of diverse genes, but diversities in these features due to environmental action are relatively rare.

This is the case also for such characters in the vertebrates, including man. Most of the physical characters of human beings form of their features, complexion, programmes of hair, form of limbs, and the like, as well as sex are settled mainly while the child is carried in the mother's body, when the differences in environment are very slight.

The diversities between individuals in these respects are almost entirely though not quite entirely the result of differences in the genes; they are matters mainly of inheritance. This is demonstrated by what we find in twins. The one-egg twins, having the same genes, are usually closely alike in their physical peculiarities, while the twins derived from different eggs, and so having different genes, are much more diverse with respect to such physical characteristics.

At times, of course, one-egg twins become diverse through accidents or diseases which affect one and not the other. But physical diversities in twins due to genetic differences are much more frequent than those due to environmental differences.

In some organisms the gross physical features are readily altered by environmental diversities, so that differences between individuals resulting from the different conditions under which they have lived are common. Such is the case with the habit of growth in higher plants; trees and other plants grown under different conditions are very different in form. In some of the lower animals, such as the hydroids, a similar situation is found.

It is obvious that the details of behaviour are in a high degree under the influence of environmental conditions. The particular acts performed at given moments in any given organism are determined almost entirely by the conditions under which it finds itself. Yet organisms of different genetic constitution behave very differently under effectively the same conditions, and the general pattern of the behaviour of organisms may be largely determined by the genetic constitution.

GENES AND ENVIRONMENT IN RELATION TO MENTAL CHARACTERISTICS

The relative rdles of heredity and environment are of special interest in man, particularly in relation to the characteristics that influence behaviour. For the study of these matters man is the most favourable organism that exists, although for most other relations in genetics he is a singularly unfavourable object of study. In man it is possible to make detailed studies of temperament, mentality, character, a group of characteristics hardly open to examination in other organisms.

There occur in man 'identical' or one-egg twins, individuals having the same set of genes, the same genetic constitution throughout. And for comparison with these there occur fraternal twins: two individuals of the same parentage, the same age and living under the same environment, but developed from two separately formed eggs and thus having to some extent different sets of genes. By comparing these two kinds of twins with one another and with unrelated individuals, opportunity is presented for comparisons of the results of similarities and diversities in genetic constitutions with those of similarities and diversities in environmental influences.

As seen earlier the twins derived from a single egg are as a rule very closely alike in most physical respects; it is this that has caused them to be called identical twins. In some cases, however, the two differ in certain physical characteristics. There appear to be two categories of causes for these differences. First, there are physical differences that result from the way in which the separation of the twins occurred. Separation of the single egg into two does not occur in the first stages of development, as by division of the one-cell stage into two. On the contrary, it occurs much later, when many cells are present and the differentiation of bodily parts has begun, and it occurs at different ages in different cases.

There is reason to believe that when an equal division occurs at a rather early stage, before the right and left halves of the body have begun to show their characteristic diversity, the two twins produced are closely alike in their physical features.

At a later period the right and left halves of the embryo have begun to differ, in the way that results as a rule in the greater strength and development of the right hand and arm. At such a stage, the embryo may divide into right and left halves, with results as follows.

Of the two embryos so produced, one will have a right limb that is already somewhat advanced in development, while the left limb, formed at the plane of separation, must begin anew. The other embryo will be in the reverse situation; it will have a left limb already somewhat advanced in development, but the right limb, formed at the plane of separation, must begin anew.

In each case the more advanced limb retains its advantage and becomes the one that after birth is strongest and most used. The result necessarily is that one of the twins remains right-handed, while the other becomes left-handed. This situation is not uncommonly found in identical twins. Similar differences in other unsymmetrical parts may be produced in the same way: for example in the whorls of hair on the head. In such respects the two members of a pair often show 'mirror-imaging', presumably due in each case to the fact that the original embryo had already become a little unsymmetrical before division occurred.

From this same situation another condition often observed is believed to result. In the embryo before division, one side (usually the right) will be a little in advance of the other. After division, the twin derived from the right half continues to retain this advantage, so that it may be more vigorous than the one derived from the left half. Such differences in vigour are not rare in twins.

Apparently they may be accompanied by a psychological difference; the more vigorous twin assumes the leadership in their lives together. This reacts further on temperament and character. How far psychological and temperamental differences so produced may go is uncertain.

It appears that in some cases division of an embryo may be unequal, so that for this reason also one twin may be more vigorous than the other, with resulting psychological differences.

Diversities produced in the ways above described are different in origin and type from those commonly called environmental. Under the latter term are commonly grouped diversities resulting from the different conditions under which individuals have lived. Since the extent to which differences due to the time and method of division may go, and the kind of characteristics that they may affect are uncertain, it is often not possible to distinguish these certainly from diversities due to differences of environment after division has occurred.

Belonging to one or the other class, there are sometimes marked differences in the phsyical characteristics of one-egg twins. Komai and Fukuoka have described two fifteen-year-old Japanese boys that are obviously one-egg twins; they show the usual close resemblance in form, features, colouring, and the like. But they now differ greatly in size.

One is taller by 14-8 centimetres, or about 6 inches, and his weight is greater by 10-4 kilograms, or about 23 pounds. The two were at first of the same size, the difference gradually arising during growth. The smaller of the twins is affected by diabetes insipidus; this is probably connected with his decreased growth. The larger twin is right-handed, the smaller left- handed. Whether the difference in health and growth is in some way connected with what occurred at division of the egg is uncertain.

A comparable case is described by Siemens. Two sisters, clearly one-egg twins, were as usual nearly identical in most of their physical characteristics till the age of 10. At this time one of them became affected wth a severe lateral curvature of the spine, and from this time their development was very different; at 16 the healthy twin was 4-8 inches taller than the other. The origin of the defect that produced curvature of the spine and changed development is unknown.

Identical twins that have lived together sometimes show considerable psychological differences. Newman has made extensive psychological tests on fifty pairs of identical or one-egg twins, and, for comparison with these, on fifty pairs of fraternal twins. The results of these examinations have not yet been published in full, but certain data from them have been published.

By the Stanford-Binet tests, the intelligence quotients (IQJ and the 'mental age' were determined for the two sets. The two classes of twins are best compared by the differences between the scores made by the two members of each pair. Thus the average difference between fraternal twins, having somewhat diverse genetic constitutions, is nearly twice as great as that between identical twins, in which the genetic constitutions are alike.

But it is important that the identical twins that have lived together also showed differences between members of the pairs. In different pairs there was much variation in the amount of diversity. Of the 50 pairs there were five in which the two members gave identical scores in the intelligence tests. At the other end of the scale were five pairs that showed respectively the following differences (points in the Stanford-Binet scale) between the two members: 12-6, 12-9, 13-0, 13-9, 16-0.

All these five differences are considerably greater than the average difference between the genetically unlike fraternal twins. It is clear that in some cases even identical twins living together differ much in mentality. Whether the differences are the result of original diversities consequent on the method of division of the egg, or have been produced in some way by different experiences of the two, is not known.

For comparisons as to the relative effects of genetic and environmental diversities on mental characteristics, valuable data have been obtained from the examination of identical twins that have lived apart under different environments. The first case of this kind was carefully studied by Muller, and nine pairs have since been fully examined by Newman. In each of these ten cases the twins were separated in infancy, being adopted into different families; they have then lived to adult life under diverse conditions and influences. In addition to careful observations as to character and temperament, they were subjected to psychological and temperamental tests by the best standardized methods available.

The type of results reached by these studies will best be appreciated by first looking at the detailed resemblances and differences in some particular cases. In the first such study, by Muller, there were twin sisters that had been separated when two weeks old, and that did not see each other till they reached the age of 18; from that time until the age of 30 they lived apart more than nine-tenths of the time.

Physically they showed the extreme similarity in characteristics that is usual in identical twins. Both 'have always been intellectually active', 'both have been extremely energetic, capable and popular, and they have been prominent in all sorts of club work in their respective communities' (Muller)'.

Both have had two or three attacks of tuberculosis, almost simultaneously'. The usual intelligence tests gave results very closely alike for the two twins. But 'the nonintellectual testsýÿof motor reaction time, association time, "will temperament", emotions and social attitudesýÿgave results in striking contrast with those of the intelligence tests, in that the twins gave markedly different scores in all these tests'.

The differences were on the average greater than those between two individuals taken at random, and seemed 'to be correlated with salient differences in their past experiences and habits of life'. Thus this first study of such a case indicated that the different environments and experiences of the two individuals had produced a large effect on temperament, emotions, and social attitudes, but had had little effect on such matters as are tested by intelligence tests.

An illuminating contrast with these results is given by the first pair studied by Newman. The twin sisters ('O' and 'A') were born in London and were separated at the age of 18 months. One lived in Ontario, Canada, the other in London. Their environments were very different. When they were tested, this pair gave differences in those tests in which Muller's pair gave similarities, and gave similarities in those tests in which Muller's pair gave differences. Newman says: 'The twins dealt with in this paper are very different in mental capacity. But they showed great similarity in their manifestations of will and temperament, and in their emotional reactions. So this pair of twins shows that differences in the experiences undergone affect deeply the individual's mental traits: his performance in matters brought out by intelligence tests.

Important points were brought out in the study of Newman's second pair, twin sisters. The two had received very different educations; one had attended school seven years longer than the other. Newman summarizes as follows the results of his study of these:

'These twins, remarkably similar after being separated at 18 months of age and unknown to each other for 19 years, have been profoundly modified by the very different educational careers.In every test of mental capacity, whether of so-called native ability or of achievement, 'G', the more highly educated twin, has distinctly the superior mind. Obviously mental training improves the ability of an individual to score well in any sort of test'.

But further, 'in contrast with the great difference in mental power stands the fact that in all the tests of emotional traits and of temperament the twins gave the impression of being remarkably and unusually similar'. Here came out the effect of their identity of genes.

Newman's third pair revealed certain other facts of importance. These were two young men ('G' and 'O'), separated at two months of age. One had lived mainly in the city while the other had lived in the country. They were examined after reaching the age of 23.

In native ability they seem to be nearly identical. The one outstanding difference is in their general personalities. 'G' (who had lived in the city) impresses one as more dignified, more reserved, more self-contained, more unafraid, more experienced, and less friendly. He seldom smiles, has a more serious expression about the brows, eyes and mouth. He stands more erectly with chin held in and brows drawn down somewhat over his eyes. 'O' (the other twin) is the opposite in all these respects. He is the more typical country boy, laughs readily, and is not on his dignity at all. Newman asserts emphatically that the personalities of the boys were utterly different.

Thus through the study of these first four pairs examined, of identical twins that have lived apart, it became clear that such twins may differ in many ways. It appeared that the different environments and experiences of the individuals may have a large effect on mental and temperamental characteristics. The effects were different in ctifferent cases: in some cases the twins are alike in intelligence but differ temperamentally; in other cases they are alike in temperament and emotions, but differ in mentality. In other cases the marked difference is best expressed rather vaguely as a diversity in 'personality'.

The general impression given-by study of these first four cases has been confirmed by Newman's study of six other pairs of identical twins that have lived apart. The nine cases examined by Newman were treated by uniform methods, so that they may be compared in such a way as to bring out general relations.

The mean difference in intelligence quotient for these nine pairs of identical twins that have lived apart is 8'6. This is considerably greater than the mean for identical twins that have lived together, which was 5-3. So far as it goes it indicates that difference of environment has a considerable effect in increasing the mental difference between twins.

In cases I, II, IV and VIII, the difference between the twins was very marked, but is similar to the differences between the five extreme cases of identical twins that had lived together. This emphasizes the fact that it is not possible to be certain what differences are due to something that happened at the division of the egg, and what are due to later environmental differences. The mean difference between identical twins that have lived apart is nearly the same as that for fraternal twins that have lived together.

In addition to the studies of intelligence, Muller and Newman made extensive examinations designed to test other features of personality. One set was designed to test will and temperament, to distinguish the mentally quick and slow, the deliberate and careful as compared with the hasty and careless, the aggressive and forceful or the opposite, and the like. Another type of examination is designed to test emotional peculiarities: likes and dislikes, feelings of right and wrong, and so on.

The results of such tests are not readily expressible by a numerical score, such as is employed for the intelligence tests. It will be worth while, however, to attempt a tabulation in general terms of the results of the tests in the three categories for the 10 cases of identical twins that have lived apart. Thus the general picture is that identical twins that have lived apart show many different combinations of likeness and unlikeness with respect to intelligence, temperament and emotional characteristics.

There are cases in which the two are closely alike in all three categories cases in which the twins are distinctly diverse in all three categories and cases in which they are alike in two categories and diverse in the third, or alike in one category and diverse in the other two.

On the whole, it is clear that other things, beside genetic constitution, play important roles in determining mentality, temperament and emotional traits. The principal difficulty in interpreting the results lies in the uncertainty as to how much is the result of the manner in which division of the egg occurred, and how much is due to later environmental differences. Both certainly play a role. For these types of characteristics it appears probable that, of the two, the later environmental differences play the greater role. It is further clear that genetic resemblances and differences play a very large role in mentality, temperament and emotional traits.

It is to be noted that in all this study, the general

environment of the twins was much alike in all cases. None of them lived in different civilizations, or at different epochs in cultural development. How great a diversity in the characteristics studied could be made by such differences remains uncertain.

It is further to be noted that the tests to which the twins were subjected were designed to bring out rather permanent personal traits, as distinguished from mental content and habitudes. The behaviour of human beings depends very largely upon acquired mental content, upon knowledge, and upon habitudes and these things are in a high degree dependent on the conditions under which the individuals live and the experiences to which they are subjected.

In general, it is clear that in a human population there are great numbers of different types of individuals that are diverse because their genetic constitutions are diverse. In consequence of this they differ in their capabilities, in their tastes, in their tendencies toward any particular line of action. But all these classes, so far as they do not fall in the seriously defective groups, have marked powers of adjusting themselves to many different conditions and of following many different lines of action, depending on the conditions in which they live.

In no other organism than man is so full an analysis possible of the relative roles of genetic constitution and environment in determining the mental and temperamental characteristics, for in no other organism are these characteristics highly developed.

Chapter 3

The Chromosomes

All Living Organisms consist of cells. Bacteria and many primitive "microorganisms" consist of single cells higher organisms are composed of vast numbers of cells. With rare exceptions, cells are exceedingly small and cannot be seen without the help of a microscope. It is filled with a living substance called cytoplasm. Suspended in the cytoplasm you see a more compact body this is the nucleus.

Every cell contains a nucleus, and although cells may survive for quite a while without nucleus, they can no longer divide. In cells treated with certain dyes, the nucleus shows a network of stained material and usually also one or a few more densely stained clumps. This is all that can be seen of the chromosomes at this stage. When a cell divides, the membrane surrounding the nucleus breaks down and the chromosomes come out into the cytoplasm.

They have now contracted into rod shaped structures which in many cells can be seen during life and without staining, but are most clearly observed in stained cells. It is their ability to take up certain dyes which has given them their name (chromosome stained body). You can easily count that there are 12 of them, and you can also see that they differ from each other in size and shape. If you would look at more Vicia cells you would find that every one of them contains the same 12 chromosomes.

Similarly, every cell of a pea plant contains 14 chromosomes; every cell of the fruit fly Drosophila melanogaster contains 8 chromosomes every cell of a mouse contains 40 chromosomes), and every human cell 46 chromosomes. In general, number, sizes and shapes of chromosomes are characteristics of each species.

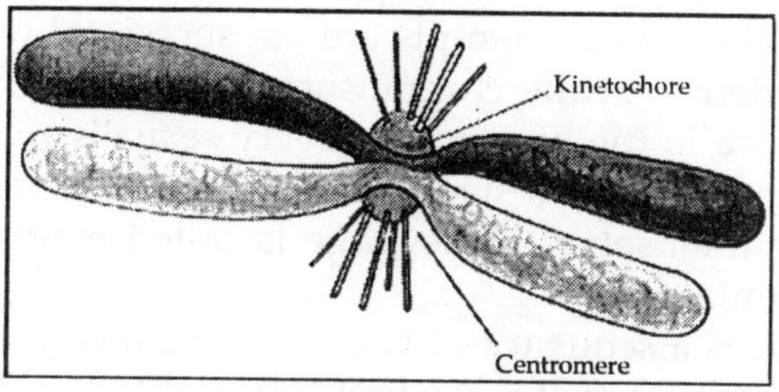

Fig. Structure of Chromosomes

How is it brought about that every cell of an individual receives the same chromosome outfit? New cells arise through division of old ones. In preparation for cell division, every chromosome makes another chromosome exactly like itself. At first these two chromosomes, the old one and the new one, lie close together; but when the cell divides they separate and each of the two sister cells receives one copy of each chromosome.

The intricate mechanism by which this distribution of sister chromosomes into sister cells is achieved is called mitosis. Since its details have no bearing on Mendel's laws. For us it is sufficient to bear in mind that, as a result of chromosome replication and mitosis, the same chromosomes are present in every cell of every organism of a given species. There is one important exception to this rule, and with this we shall deal now. It will take us right up to the point where genetics joins hand with cytology, the study of the cell.

Let us turn once more to the photographs. Cells with few chromosomes you will notice that these can be arranged into pairs. Every chromosome that stands out by its size and shape can be matched by another one of the same kind. Thus in Drosophila, we find one pair of medium sized chromosomes, two pairs of long ones, and one pair of very small, dot like chromosomes.

In cells with many small chromosomes the arrangement into pairs may not be obvious. The human chromosomes) can be arranged fairly convincingly into pairs by cutting out the photographs of all individual chromosomes and matching them according to size and shape.

Even where such attempts are not successful we can feel quite confident that the chromosomes can be arranged into pairs, because in the life cycle of every sexually reproducing organism there occurs a stage when the chromosomes do the pairing off themselves. This stage is called meiosis, which signifies diminution.

Meiosis is a sequence of two cell divisions which follow closely on each other; it is the first step in the formation of the mature germ cells or gametes. A grossly oversimplified diagram of meiosis for one pair of chromosomes; it shows only those features that are relevant for an understanding of Mendel's laws. At the beginning of meiosis partner chromosomes come together and wind around each other in various ways. As in many old dances, pairing of partners is the prelude to their moving apart.

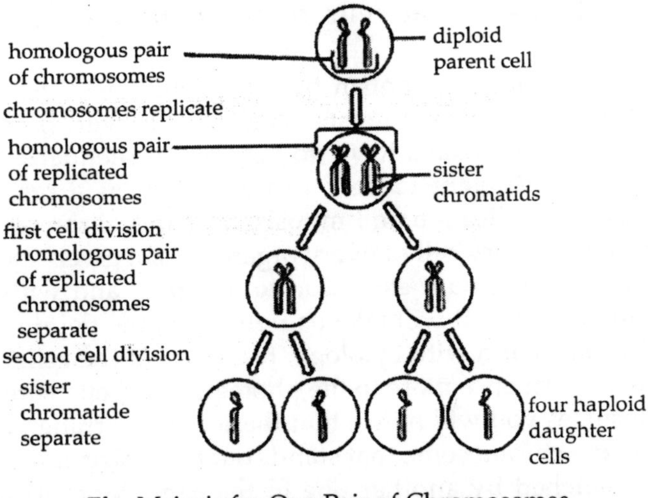

Fig. Meiosis for One Pair of Chromosomes

At the end of the first meiotic division, all pairs have dissolved, and the partners have moved separately into the two sister cells. Again, as in mitosis, division has resulted in an equal distribution of chromosomes into sister cells, but there is an all important difference. In mitosis, each cell receives the full complement of chromosomes; in meiosis it receives only one member of each chromosome pair. Thus, the first meiotic division reduces the chromosome number to half; it is therefore called the reduction division. The second meiotic division is an ordinary mitosis and does not produce any more changes in chromosome number.

The final outcome of the two meiotic divisions is a group of four germ cells, each of which contains one "set" of chromosomes, that is, one member of each chromosome pair. Such a cell is called "haploid," while cells with two matching chromosome sets are called "diploid." Expressed in these terms, meiosis produces four haploid germ cells from one diploid mother cell.

The chromosomes in an ovum (egg) of the mouse; you may count that there are 20 of them, half as many as in an ordinary body cell of the mouse). At fertilization two gametes, each carrying one set of chromosomes, come together and restore the diploid condition. The circle is closed, and a new circle will start when the next generation forms its germ cells.

THE GENES

The genes are much too small to be seen individually, but in certain stages many chromosomes show lengthwise patterns of knobs or bands that indicate their subdivision into genes. In par ticular, this happens during the early stages of meiosis. A pair of rye chromosomes at the beginning of meiosis. You notice the pattern of darkly stained knobs and faintly stained connecting strands.

The pattern in the two partner chromosomes is the same, and pairing is exceedingly accurate, each knob in one chromosome lying next to the matching knob in its partner. It appears that it is the genes rather than the chromosomes as a whole that direct the pairing process. This ability of the genes to find and attract their exact counterparts is one of their many remarkable properties for which science has not yet found a satisfactory explanation.

Later on in meiosis the pattern will be lost because the

chromosomes curl up and contract into the compact bodies. But when the next generation makes germ cells, each chromosome will again show its characteristic pattern, so that a trained cytologist can recognize every one of the seven chromosomes of rye by the pattern of its knobs.

Giant chromosomes with beautiful and intricate patterns of dark and light bands are found in the salivary glands of the larvae (maggots) of certain flies. Chromosomes in the salivary glands of Drosophila larvae. They are hundreds of times as long as ordinary chromosomes, and each of them has its own characteristic pattern of bands.

They are also much thicker than ordinary chromosomes because each consists of a bundle of many strands, all with exactly the same pattern of bands. The way these bundles arise is interesting because it illustrates the extreme accuracy of chromosome replication. In the young larva, the chromosomes in the salivary glands are still thin. As the larva grows, the salivary glands grow too, but they do so in a particular way.

The cells, instead of making more cells by division, grow bigger and bigger without division. Their nuclei, too, grow bigger and bigger. Inside the nuclei, the chromosomes make replicas of themselves as they do in preparation for mitosis but, as there is no mitosis, the new chromosomes do not separate from the old ones. Instead, new and old chromosomes remain closely attached to each other and each, in turn, makes more replicas.

This goes on repeatedly, and so accurately does each band make another band like itself that the whole bundle appears as one single thick banded thread. In addition, the partner chromosomes in the salivary glands are closely paired as in meiosis, so that each apparent chromosome represents one pair, and the nuclei of the salivary glands of Drosophila seem to have four chromosomes instead of the eight that are visible in ordinary cells).

The pattern of bands corresponds to the pattern of underlying genes. This can often be seen in cases where a chromosome has lost a small piece through radiation or some other injury. The accuracy of chromosome replication is such that all descendants of the damaged chromosome lack exactly the same piece, and when such a damaged chromosome gets into a gamete, for instance an ovum, individuals are formed which lack the same chromosome piece in all their cells.

The left part of a normal wing of Drosophila and above it, and at a very much higher magnification, a little section of one of the normal salivary gland chromosomes. The arrow points to a band which encloses a gene for wing shape. The right part of the same chromosome section in a fly which had received a damaged chromosome from its X-rayed father or mother.

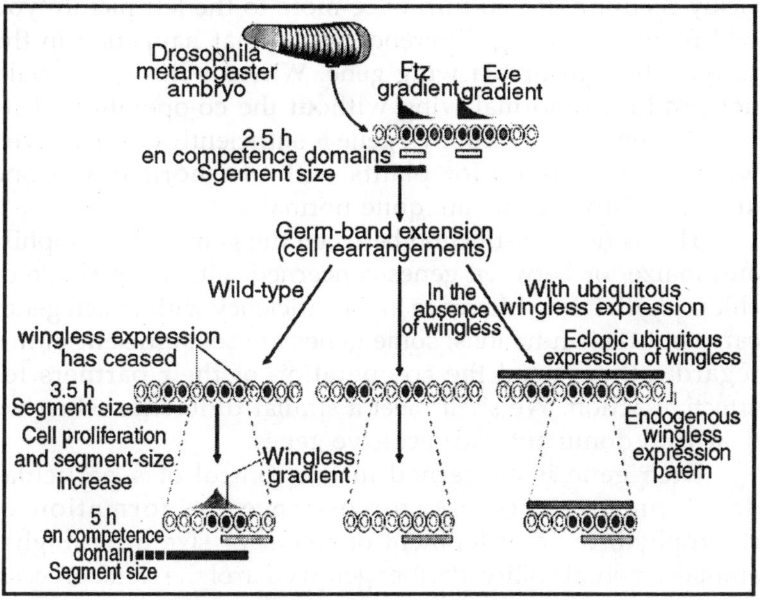

Fig. Wing Gene in Drosophila

If you recall that the salivary gland chromosomes consist of two chromosome bundles formed by the two partner chromosomes, you will understand why only half a band is missing at the point of the arrow. Nevertheless, the lack of one of the two partner genes has been sufficient to disturb normal development of the wing and to produce the abnormal shape shown in the accompanying sketch. Flies which have lost this band from both partner chromosomes are so much damaged that they die as early embryos.

The left drawing represents a normally growing green maize plant and underneath, at much higher magnification, the tips of one particular chromosome pair. There is a fairly large knob at the end of these chromosomes. Among the genes which this knob encloses is one whose action is necessary for the production of the green substance chlorophyll that is essential for survival of the plant.

The right picture shows that, when this knob has been lost from both chromosomes, the plant dies early as a yellow and sickly seedling. If you turn once more to the left picture you will notice a striking difference from what happened in the case of the Drosophila wing gene. Whereas this gene could not produce a normal wing without the co-operation of its partner gene, the chlorophyll gene is sufficiently effective even when acting singly; for plants with one normal and one knobless chromosome are quite normal.

This is not a distinction between the genes of Drosophila and maize, or between genes concerned with wing shape or chlorophyll. It is a difference in the efficiency with which genes carry out their functions; some genes are selfsufficient in this regard, others need the co-operation of their partners for effective action. We shall meet a similar difference when we deal with dominant and recessive genes.

Each gene is concerned in the control of a particular developmental process, for instance the formation of chlorophyll, the development of colour or size or, in higher animals, mental ability. Partner genes control the same process, but they may do so in different ways; thus two partner genes in peas are concerned with the control of seed shape, but one causes the seeds to be round, while the other makes them wrinkled.

Partner genes with different effects are called allelomorphs or, for short, alleles. The term "allelomorph" signifies "different shape"; according to modern ideas this may be interpreted as different chemical configurations of basically the same gene.

All developmental processes are highly complex, and it is therefore not surprising that many different gene pairs are involved in the development of, say, an eye or a brain or a flower. Even such a seemingly simple achievement as the formation of a vitamin by a bacterium proceeds in a series of steps and is controlled by a series of genes.

In bacteria, genes controlling the same process tend to lie close together, as though this facilitated their co-operation; but in most other organisms this is not found to be so. Genes which are involved in quite different processes may lie cheek by jowl on the same chromosome, while genes controlling closely related processes may lie far apart. The position of a few of the known genes on a particular pair of chromosomes of the fruit fly Drosophila melanogaster.

Only genes concerned with the development of the eyes and the wings are represented; more genes controlling development of these organs are present on the same chromosome pair and on the three others. Some of the represented genes can be seen to act in very different ways; thus wing gene w_1 is concerned with the control of wing shape, while wing gene w_2 helps make the correct pattern of wing venation.

For other genes such a distinction is not evident; thus eye genes e_1, e_2, e_3 and e_4 are all engaged in the formation of eye colour. Yet we know from developmental studies that each of them has a different role to play in this process.

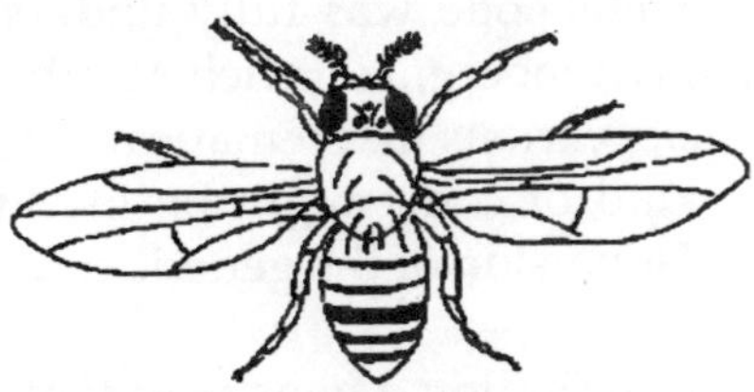

Fig. The Positions of Some Wing and Eye Genes

The genes, like the chromosomes, are arranged in pairs; with the chromosomes they segregate into different gametes at meiosis and come together again at fertilization, one member of each pair being derived from the father, one from the mother. The arrangement of the genes on the chromosomes is fixed for each species, and partner genes occupy identical positions on partner chromosomes.

Allelomorphs are partner genes with different effect, that is, they are genes which occupy the same position on the same chromosome pair and control exactly the same developmental process, but they do so in different ways.

Genetic code

Even more remarkable is that non-overlapping triples of the letters *A, C, G* and *T* which are often referred to as codons (especially when the triples are part of a gene), may spell out (reading in the 5' to 3' direction) the order of the amino acids that form the long linear molecules known as proteins. It turns out there are 20 amino acids which are the building blocks for proteins. Since there are 64 possible triples (repeated letters within the triples are allowed) using the 4 letters *A, C, G,* and *T,* different triples can represent the same amino acids.

There are also 3 triples (*TAA, TAG, TGA*) known as stop codons that do not represent (a common phrase being do not code for) an amino acid, but are involved in signaling a termination to the protein production process. One particular codon (*ATG*), which does represent one of the amino acids, is an indication for the start of the production of a protein. This system for coding proteins has come to be known as the genetic code.

Before the genetic code was fully understood, based on work by Marshall Nirenberg, Heinrich Matthaei, Har Gobind Khorana, and others, various mathematical ideas were invoked to suggest what kind of code might be involved. There is a diagram of the RNA version of the genetic code, where *U* takes the place of *T*.

One approach to finding a gene is to look for a stretch of DNA which starts with ATG (the start codon) and ends with one of the stop or termination codons. This situation in a stretch of DNA is referred to as an open reading frame (ORF). However, not all ORFs correspond to a stretch of DNA which will initiate the production of a protein, and so the location of an ORF is not equivalent to finding a gene. A major problem facing biologists is how to locate those stretches of DNA that are genes in the large amounts of DNA found in a chromosome.

The way that DNA is involved in the production of proteins is not direct. The mechanism involves another helical molecule called RNA, of which there are a variety of types. (Among the types of RNA are messenger RNA, or mRNA, ribosomal RNA, or rRNA, and transfer RNA, or tRNA.) When a protein is to be produced, the DNA separates and a copy of the gene is transcribed (the process is referred to as transcription by molecular biologists) into a strand of RNA (ribonucleic acid).

RNA, like DNA, uses an alphabet of 4 nucleotides *A, C, G,* and *U* (for uracil, a pyrimidine like thymine which takes the place of *T*). In a somewhat complex series of steps, a protein is produced. A very rough schematic of how genes create proteins. This schematic is often referred to as the central dogma.

The original DNA, which are known as introns but which are not involved in the production of a protein, are separated or snipped out and a resulting string of exons from the original DNA is left. What is left is sometimes referred to as a coding sequence, and it contains the codons (in the letters *A, C, G* and *U*) which will create the protein, by the process usually referred to as translation.

The actual manufacture of the proteins involves structures called ribosomes. Production of proteins does not go on within the nucleus but in the cytoplasm surrounding the nucleus. One can express the codons involved in the manufacture of proteins either in DNA terms (using the letters *A, C, G* and *T*) or in RNA terms (using the letters *A, C, G* and *U*).

Implications

Many interesting questions can be asked about the genetic code, including why the pattern of redundancies in the code exists. While at one time it was thought that the code was totally universal, this has proved not to be true. For certain bacteria and in other situations the universality of the code breaks down. Cellular life forms are classified according to whether they are prokaryotes or eukaryotes. Prokaryotes are cellular organisms where the genetic material is not a part of a membrane enclosed nucleus, while in eukaryotes the genetic material is contained in a nucleus.

At one time life forms were classified as either bacteria or eukaryotes. A group of bacteria like organisms, found in very hot environments such as undersea vents, were initially classified as bacteria. However, when Archaea's genetic makeup was studied, its members were found to have an amazingly large number of genes that were not known to exist in other organisms.

They seemed to be surprisingly different from the bacteria with which they had been formerly classified. It has become standard in recent years to list the Archaea in a separate category from the bacteria and eukaryotic organisms. Use of phylogentic trees was in part a tool that led to this change.

It is tempting to believe that, because human beings are as complex as they are, other species will have less genetic material than humans. However, chromosome number (and the total size of the genome) varies with species in a way that seems to have no rhyme or reason. Thus, chimpanzees have 48 chromosomes, cats 38, dogs 78, rabbits 44, rats 42, turkeys 82, and horses 64. Perhaps when the genomes of many different and varied species are sequenced it will be possible to get insight into these wildly varying numbers.

MENDEL'S FIRST LAW: SEGREGATION

After this excursion into cytology, lot us return to genetics. The chief tool of the cytologist is the microscope; it has shown him that the chromosomes, and with them the genes, segregate at meiosis and enter the gametes singly. The chief tool of the geneticist is the crossing of different strains, and we are now going to consider crosses which demonstrate the segregation of the genes at meiosis.

It will be useful for the description of these experiments if we recall some technical terms which have already been defined, namely "haploid" and "diploid" for cells or individuals which carry one or two chromosome sets, respectively, and "allelomorphs" or "alleles" for partner genes with different effects. To these terms I want to add two new ones which describe the genetical constitution of a diploid individual in regard to any particular pair of genes.

Evidently there exist two possibilities: either the partner genes are alike or they are allelomorphic. In the first case, the individual is "homozygous" or a "homozygote" in respect to the gene concerned; in the second case, it is called "heterozygous" or a "heterozygote." A pea plant with two genes for wrinkled seeds or two genes for round seeds is homozygous for either wrinkled or round; a pea plant with one gene for wrinkled seed and an allelomorphic gene for round seed is heterozygous for wrinkled and round. It will readily be seen that only heterozygotes are suitable for demonstrating the segregation of genes; for the geneticist recog nizes genes by their effects, and the segregation of like genes in a homozygote is not observable.

In higher organisms (macroorganisms) it is hardly ever possible to observe the immediate effects of segregation at meiosis; for the mature gametes fuse at fertilization and the progeny, in which we observe the effects of the genes, carries genes from two different gametes

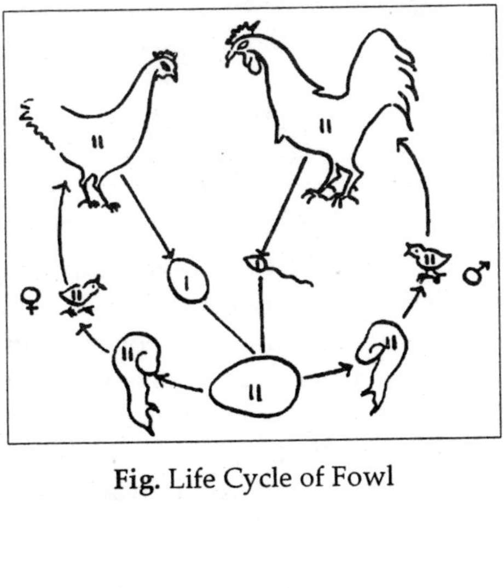

Fig. Life Cycle of Fowl

The life cycle of the domestic fowl in terms of haploidy and diploidy. The diploid female and male form haploid ova and spermatozoa. At fertilization the gametes fuse and form the fertilized ovum, which in turn develops into the diploid embryo, chicken and adult. Thus the whole life cycle, with the exception of one cell generation, is spent in the diploid stage.

The situation is quite different for many microorganisms. The life cycle of a green alga, Chlamydomonas. Each individual of this species is a microscopically small cell which swims through the water with the aid of two whiplike hairs, or "flagellae." These cells are haploid; they have only one set of chromosomes, carrying all genes of the species in single copies. New cells are formed through division of the old ones.

In this way whole populations may be derived from one cell by repeated divisions. Such populations are called "clones." Since at mitosis both daughter cells receive exactly the same set of chromosomes and genes all cells within a clone are genetically identical.

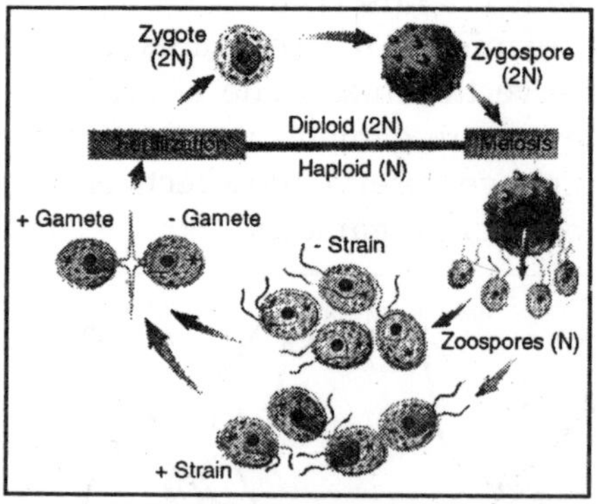

Fig. Life Cycle Algae Chlamydomonas

There are no visibly distinct sexes, but physiologically there is a clear distinction between sexes or "mating types," called + and – in our diagram. Cells of the same clone are always of the same mating type and never mate with each other; nor does mating occur when cells of two + clones or two – clones are brought together.

But when cells from *a*+ and *a*– strain are put into the same culture vessel, pairs are fomed. Each pair consists of one + and one – cell they fuse and form a thick skinned spore called a zygote. The zygote is comparable to a fertilized ovum in, say, a dog or a fly. It contains the chromosomes from both gametes and is, therefore, diploid. Very soon after fertilization meiosis takes place inside the zygote, and haploid gametes swarm out.

These start a new life cycle by division; in fact, the gametes of Chlamydomonas do not differ from any of the other free swimming cells. Thus in this algae, as in many other microorganisms, the whole life cycle, with the exception of the zygote, takes place in the haploid stage, and segregation at meiosis is not obscured by diploidy of the progeny.

Usually it is necessary to grow the gametes into clones before segregation becomes observable; but occasionally the segregating gametes themselves differ visibly from each other. Two strains (1) and (2) differ in one gene which is concerned with the development of the flagellae. While strain (I) has all the genes necessary for flagella formation, strain (2) carries a deficient allele of one of these genes and cannot, therefore, develop flagellae.

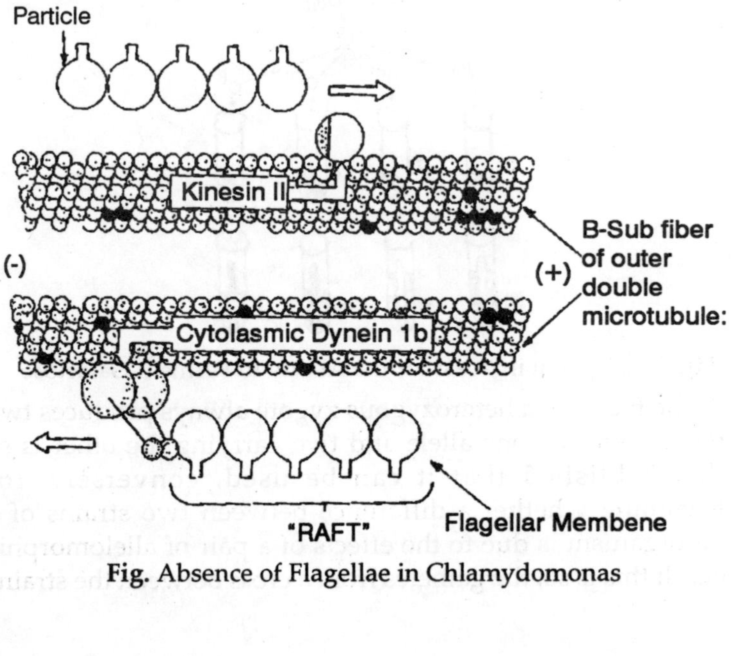

Fig. Absence of Flagellae in Chlamydomonas

When the two strains are crossed by bringing together + cells from the one strain with cells from the other (it makes no difference which strain provides the + cells, which the cells), a zygote is formed which carries both alleles and is, thus, heterozygous for the gene in question. At the reduction division the partner chromosomes segregate and carry the different alleles into separate cells. At the second meiotic division each cell divides once more by mitosis, so that finally four gametes are formed, two of which carry the normal and two the deficient allele. When these swarm out, it will be seen immediately that two have flagellae, two have not. This difference persists in the clones formed from the gametes.

If we let each gamete grow into a separate clone, two clones will consist of flagellated free moving cells, while two will consist of flagellaless cells which sink to the bottom of the tube.

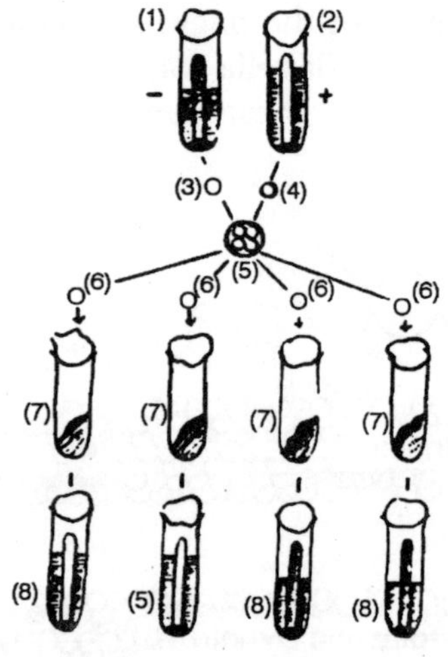

Fig. Segregation in Yeast for the Ability to Ferment Galactose

The fact that a heterozygous zygote always produces two gametes carrying one allele and two carrying the other is so well established that it can be used, conversely, for determining whether a difference between two strains of a microorganism is due to the effects of a pair of allelomorphic genes. If this is so, the gametes from a cross between the strains will yield two clones, resembling the one strain, and two resembling the other. Yeast, like Chlamydomonas, is a unicellular organism. Many strains are haploid and can be used in experiments on segregation. Like Chlamydomonas, yeast occurs in two mating types, + and –, which are not visibly different from each other.

Yeast, as you know, ferments sugar into alcohol and carbon dioxide gas; but strains differ in their ability to ferment various kinds of sugar. Thus some strains readily ferment galactose (one of the components of milk sugar), while others do so only very slowly and inefficiently.

A simple method for detecting fermentation is to catch the developing gas in a test tube which was first completely filled with liquid and then inserted immediately above the yeast cells. If the medium contains a sugar which can be fermented, the tube gradually fills with gas; if the sugar cannot be fermented, the tube remains filled with liquid.

In the experiment shown in the medium in the top row contains galactose as the only sugar. Strain (2) is able to ferment this sugar, strain (1) is not. Cells (3) and (4) from the two strains are brought together, and a zygote (5) is formed which, through meiosis, gives rise to four gametes (6). Each gamete is grown into a clone (7) on solid medium containing glucose, a sugar which can be fermented by all strains of yeast. Finally, the four clones are tested (8) in fermentation tubes with galactose as the only sugar. You will notice that two clones (those on the left) can ferment galactose, and two cannot.

Thus the difference between the two original parents segregates in the 2:2 (or 1:1) ratio expected if it is due to the action of a single gene pair. On the basis of several experiments of this kind it could be concluded that in yeast the ability to ferment galactose is controlled by the action of a gene which in nonfermenting strains is absent, or is present in the form of a nonfunctioning or poorly functioning allelomorph.

The essence of what is called Mendel's first law or the law of segregation. In modern terms it may be expressed as follows: When a heterozygote forms gametes, half of them carry one of the two alleles, half the other. The basis for the segregation of the genes is the segregation of the chromosomes at meiosis. Mendel himself could not, of course, interpret his results in these terms, but he quite clearly postulated that a mechanism of this kind was responsible for his data, in particular for the famous and much overrated 3:1 ratio. This particular ratio is a rather remote consequence of the underlying gametic segregation in a ratio of 1:1. The steps by which, in crosses between higher organisms, this 1:1 ratio becomes transformed into others, including 3:1.

In doing this we shall find ourselves using very simple statistical rules, such as are used for predicting that a coin tossed 100 times will show head about 50 times. This statistical aspect of genetics has been held against it by the Lysenko school, on the grounds that mathematics must not be applied to biological objects and that Mendel had no right to speak of a 3:1 ratio between round and wrinkled peas if some of the pods contained 6 round peas. No matter what we think about these arguments—and to a modern biologist they seem entirely futile.

Chapter 4

Genetics of Exploratory Behaviour

How Behaviours are labeled, defined, combined, and measured have a considerable amount of impact on generalizations that can be made from them and, consequently, on our understanding of their importance in the life and evolution of organisms. Although problems of definition and measurement are important in any area of Behavioural investigation, they are especially critical in Behaviour genetic analyses, for unless a Behavioural phenotype is consistently and unambiguously measured, there cannot be consistency and unambiguity regarding its genetic determinants or its relationship to other Behavioural, somatic, or morphological phenotypes.

There are few areas of Behavioural investigation in which this "phenotype problem" is so clearly shown as in the genetic analysis of activity and exploratory Behaviour directly and indirectly.

Activity and exploratory Behaviour are fundamental, yet surprisingly complex Behavioural phenotypes. Important in their own right for the organism, activity and exploration are also of importance to researchers in that each of these Behaviours has been used as an index for more abstract Behavioural concepts, such as emotionality and curiosity. We have dealt solely with the genetic determinants of activity and exploratory Behaviour.

To do so adequately we needed to go into some detail regarding definitional problems and techniques of measurement. We did not intend to provide a thorough review of the voluminous literature on these topics, but rather have emphasized studies that provide clear statements of genetic effects or of a genetic model, or that serve as definitive examples of a particular technique. An additional criterion in the selection of a major reference was to select, where possible, the study that would be most accessible to the greatest number of readers.

MEANINGS

Any movement of any part of the body of an animal is, according to a broad definition of the term, *activity*. However, as we use the term, and as it is most commonly used in the literature, "activity" refers to movement of the whole organism. In order to distinguish whole organism movement from such Behaviours as orienting response (head turning responses, movement of the pinnae, or movement of the eyes), the term *locomotor activity* is often used.

Fuller and Thompson recommend *ambulation* as a more unambiguous term. The problems with activity as a phenotype, however, derive not so much from nomenclature as from what this form of Behaviour represents, and how it is measured.

Activity as a Trait

For several decades prior to the 1960s, activity *qua* activity was outside the mainstream of American psychology. During this era of the major Behaviouristic learning theories, the concept of "Behavioural homeostasis" was in vogue. This concept held that the "normal" state of an organism was not behaving (i.e., being inactive), and that activity occurred only as the result of a "drive," that resulted from an imbalance in a physiological system, such as a tissue deficit or increase in hormonal activity.

According to the assumption of Behavioural homeostasis, then, one does not (indeed, cannot) account for activity as a general trait, because activity occurs only as a consequence of other internal and external events. In recent years, the waning influence of the major learning theories, together with a number of disparate empirical findings (for example, exploration of their surroundings by satiated animals; the neural and bodily activity seen in many species during sleep), has led to a decline in the acceptance of the Behavioural homeostasis assumption. There is an increasing degree of acceptance of its opposite: The normal state of an organism is behaving (i.e., being active).

When comparing the amount of movement of different individuals within similar settings, in the absence of specific physiological states, external goals, or noxious stimuli, one is investigating activity as a trait.

If individual organisms do indeed possess specific "general activity levels" that yield strong predictions of other sorts of Behaviour or that are highly consistent across environmental settings and resistant to alteration, the theoretical and practical implications would indeed be great. Knowledge of the extent and nature of the genetic determinants of activity as a trait would magnify the importance of these implications.

Activity as an Index

Activity has been of interest not only as a trait in its own right, but also as an index, or means of measurement, of more complex and difficult to define forms of Behaviour. The most frequent uses for activity measures in this regard are exploratory Behaviour, whiich we discuss at length and emotionality. As an index of emotionality, activity is often combined with other measures, especially urination and defecation. Stemming largely from the selection studies in rats by Hall in the early 1930s, emotionality is, in turn, an indication of the autonomic and Behavioural responsiveness of organisms to potentially dangerous or noxious situations.

Organisms are assumed to be either more active or much less active (e.g., "freezing") in these situations, depending on the species or the genotype within a species.

The advantages of using activity as an index are obvious:

- Activity is relatively easy to measure, with a number of measurement procedures yielding reliable, readily interpretable scores

- A considerable body of activity data exist for a variety of species of laboratory animals
- A number of successful selection and strain comparison studies have been completed, providing an abundance of subjects more or less "genetically standardized" on this phenotype.

The problems associated with using activity as an index are more subtle, but no less real. If activity serves as an index of a more complex characteristic, such as emotionality, there should be a strong correlation between activity and any other index measures. Such a correlation would indicate that the various phenotypes bear some causal relationship to one another; or that they are due to pleiotropic or linkage effects of the same genes; or, on the other hand, that the correlation between indices is a spurious, chance occurrence, perhaps furthered by measurement artifacts.

When the correlation between indices is high for some selected lines but not for others, or for some inbred strains but not others, one must give the greatest credence to the last of these explanations. We deal with this problem, a major factor in phenotype confusability.

When activity is used as the sole index of a more complex phenotype, it is the same as operationally defining the more complex phenotype in terms of the measure of activity (for example, defining exploratory Behaviour *as* "squares crossed in an open field"). It then becomes the obligation of the "consumers" of such information to decide for themselves whether this sort of measure is appropriately definitive and adequately representative of the complex phenotype.

An example of the problems that can arise from the use of activity as an index can be illustrated by a study performed by Whimbey and Denenberg, and further discussed by Denenberg in 1969. In the portions of this study relevant to our discussion, they placed rats in an open field briefly for 5 days and obtained a variety of measures that were then intercorrelated and factor analyzed.

Correlated positively with defecation scores and could be

considered an index of "emotionality." However, on Day 2 and thereafter this same measure of activity, taken under presumably the same conditions from the same subjects, produced a *negative* correlation with defecation and yielded different factor loadings. Then, the authors imply, activity becomes an index of exploratory Behaviour. Is activity a useful index for more complex Behavioural phenotypes?

MEASUREMENT

The Activity Wheel

This simple, time honored device is still available from a number of laboratory equipment suppliers in sizes suitable for rats, mice, and other small animals. These devices usually operate a counter that provides an easily understood and interpreted measure of activity expressed as the number of revolutions per unit of time. (Most commercial activity wheels will run both clockwise and counterclockwise, and will operate the counter when turned in either direction.) In addition to obvous extraneous variables such as room temperature, illumination level, and the like, the major variables to be controlled with this type of apparatus are:

- The degree of effort needed from the subject to overcome friction in order to turn the wheel; and
- The degree of choice available to the animal to move freely from an adjoining cage into the wheel. Assuming the greatest possible control, the activity wheel provides what is, perhaps, the purest measure of ambulation, at least in terms of energy output. However, the very conditions providing the control and "purity" of this measurement technique also make it the most artificial; ordinarily, the movement of an organism results in a changing environment for the animal as it progresses. Inasmuch as running in the wheel does not lead to changes in the animal's environment and does not therefore result in exploratory or neophobic excursions, it is not surprising that the "distance" run in such a device is

often greater than the distance traversed under more natural conditions. Perhaps this type of activity, unlike that measured by other means, is primarily exercise.

Maze Running

Where the activity wheel allows only totally circumscribed locomotion, the maze allows some degree of directionality of movement. The maze thus measures ambulation in a relatively more naturalistic fashion than does the activity wheel, especially for thigmotaxic and burrowing species, such as rats. However, it is possible that an animal moving through an unfamiliar maze is also exploring its environment, or seeking an escape route, or is engaged in some other form of Behaviour that involves ambulation. Therefore, attempting to measure *activity* in a more realistic manner can create questions as to just what is being measured.

Activity Meters

These are automated measuring devices that unobtrusively detect and count movements of animals. Newer versions consist of a platform containing a series of electromagnetic sensors upon which any nonmetallic container with one or more animals inside can be placed. Any movement by the animal operates a counter, with the degree of movement required for detection controllable by setting the sensitivity of the sensors.

The activity meter has the unique property of permitting measurement of movement in any sort of box or container that will fit on its surface, including home cages. Another unique property of this type of device is that it measures all movement by the subject, rather than merely ambulation from one location to another.

Although this has the advantage of being a meaningful measure of activity as a trait, it dictates caution when comparing scores with those obtained in other types of activity measuring devices. The unique and specific conditions that must be controlled when using an activity meter involve the placement of the animal container over the sensors and the sensitivity setting of the meter.

Although the raw scores taken from the counter activated by movement over the sensors are difficult to interpret, meaningful and statistically reliable ordinal data have been obtained with this device in studies comparing genotypes and experimental conditions.

The Open Field

This is the most frequently used technique for the measurement of activity, a fact that should have resulted in a major advantage for its use: the establishment of reliable standardized scores to serve as a basis for comparison across genotypes and experimental conditions. There is also a high degree of face validity for its use in the measurement of activity as a trait—if ambulation within an open arena is not activity, what is?

The open field can be constructed in a variety of shapes (round, square, and rectangular being the most popular) and made in sizes appropriate for almost any species. More limited versions of this apparatus are the straight runway and the shuttle box, which share many of the advantages and disadvantages found with the open field.

Unfortunately, despite its apparent advantages, the open field test has not turned out to be free of severe problems. As Walsh and Cummins point out in their thorough and critical review of the many studies using this technique: "Apparatus, techniques, subjects, parameters, analyses, and interpretations have diversified enormously, yet in the literature, sweeping generalizations and conflicting interpretations continue to be made on the basis of univariate studies lists the variety of Behavioural and autonomic measures that have been taken, although number of squares crossed seems to be the most common measure in most Behaviour genetic studies. One of the difficulties referred to by Walsh and Cummins results from the attempt to compare studies that have used different measures.

Many independent and dependent variables have been

tested, and there are numerous and sometimes inconsistent effects due to several conditions inherent in the testing situation, such as size of the open field, level of illumination, or prior experience in the apparatus. When genetic studies are reported in which such conditions are not standardized or systematically varied, it is difficult to see how valid comparisons and generalizations can be made.

Table. Classification of Open Field Dependent Parameters

I. Behavior		
	A. *Whole or major body movement*	
	1. Type of movement	
		a. Distance covered per unit time
		b. Time spent in ambulation
		c. Rearing frequency
		d. Escape attempts
		e. Latency (usually time taken to leave start area)
		f. Time spent without movement
	2. Locations	
		a. Field area visited (inner or peripheral areas, corners, etc.)
		b. Affiliation (distance from partner subject)
		c. Stimulus interaction (e.g., distance from stimulus object)
	B. *Part body movement*	
		a. Manipulation of objects
		b. Sniffing
		c. Scratching
		d. Digging
		e. Teeth chattering
		f. Grooming
		g. Vocalization
		h. Visual exploration
II. Autonomic nervous system		
		a. Defecation
		b. Digestive transit time
		c. Urination
		d. Heart rate and rhythm
		e. Respiratory rate

Despite its high face validity, ease of scoring, and long history of usage, the sensitivity of subjects to changes in experimental conditions in open field testing suggest that this technique is not the *sine qua non* for the determination of "general" activity levels. Because there are a number of questions as to just what is being measured in the open field, we suggest that studies using open field measures (especially those using a single dependent variable) be interpreted with caution.

GENETIC DIFFERENCES

The presence of a genetic effect was originally established for activity through selective breeding of animals showing high activity scores, low activity scores, or both on a standardized measuring technique. One of the earliest attempts to establish the hereditary basis of activity using this method was a selection study by Rundquist, in which he selected rats for high and low activity following testing in an activity wheel.

His results were disappointing initially, because of an error in the method of selection he employed: the mating of high active and low active offspring from nonselected parents. That is, rats showing high activity scores were selected not only from parents that also had high activity scores, but from those yielding low activity scores as well, with the reverse holding true for rats with low activity scores. As a result, there was no differentiation between groups from generations F_1 to F_4. When this error was corrected, differentiation into high and low active groups did occur and laid the foundation for further research in this area.

Hall performed selection experiments in the early 1930s to establish the genetic basis of emotionality in rats using as phenotypes the Behaviour of the rats in an open field . Although he was not concerned with activity per se, his selection procedures and his use of the open field test have served as the prototype for many of the modem studies on this problem.

In the 1950s, it became possible to use another method for determining hereditary basis for activity and for any other phenotype. This method involves finding differences in activity between strains of genetically controlled animals, usually highly inbred strains of mice and rats, but also established breeds of dogs and inbred strains of various species of *Drosophila.*

Thompson tested five strains of inbred mice on two activity measures, the open field and the Y maze. His results showed marked differences in the activity levels of the five strains, and the two C_{57} strains showed much higher activity levels in both open field test and the Y maze than did the BALB/c mice.

In mouse studies, McClearn has shown that there is a clear relationship between locomotor activity and the percentage of C_{57}BL/Crgl genes present (from backcrosses with A/Jax mice). Thompson illustrated that there is considerable variability between strains in exploratory activity, with C_{57}BR/a, C_{57}BL/6, and C_{57}BL/10 having the highest mean score (for activity) and A, AK/e, and BALB/c strains having the lowest.

These orderings were fairly invariant with another testing apparatus (enclosed arena versus Y maze). There are differences in the ordering of mice on exploratory activity with different environmental conditions. For example, he found that in bright white light, A/Crgl mice showed less exploratory activity than in dim red light. In the C_{57}BL/Crgl mice, activity increased in white light and decreased in dim red light. Although he found no change in the ordering of the strains, McClearn did find a significant interaction between strain and illumination.

Of particular historic and current interest are the MR and MNR rats originated by Broadhurst. Using a procedure essentially similar to that used by Hall more than two decades earlier for selection for emotionality, Broadhurst made several important improvements; the open field was more uniformly and brightly lighted and initial responses of the subjects were weighted more heavily, giving less emphasis to the possible adaptation of the rats to a familiar environment. MR and MNR rats clearly differ on open field defecation scores, an index of emotionality, and are often regarded as "benchmark" strains for this category of Behaviour. However, on several *activity* measures, these strains do not differ greatly. Selection studies illustrating genetic differences in activity have also been carried out in species other than rats and mice.

The comparison of the normal Behaviour of different established breeds of dogs provides an excellent example; for "everyday" differences in activity levels, one need only compare a bassett hound and a fox terrier. Dogs have also been used as subjects in more systematic genetic studies, with different breeds treated analogously to inbred strains of rodents.

A prime example of this sort of research is the 19 year study conducted by Scott and Fuller that dealt with, among other things, the emotional Behaviour of several breeds of dogs, together with their hybrids and backcrosses. Clear genetic differences were found in the five breeds on a variety of reactivity tests, including open field Behaviour.

Some work on activity and its implications has also been done with the fruit fly. In one study flies were selected for mating speed. Activity differences for these two selected strains (fast mating speed and slow mating speed) were measured by counting the number of squares a fly entered in an open field arena over a given period of time. It was found that slow mating lines exhibited more open field activity than did fast mating lines.

In contrast, Ewing, in selecting for spontaneous activity, found that the less active lines displayed *more* sexual Behaviour than did the more active lines. Although these results appear to contradict Manning's findings, the discrepancy appears to have been due to an "apparatus effect," as he used an apparatus other than the open field used by Manning. When Ewing replicated his study using the type of open field apparatus used by Manning, no significant differences between the selected lines were found.

Besides being alterable by apparatus effects and other environmental changes, activity is readily changed, often differentially for different genotypes, by internal events, especially pharmacological agents. A thorough review of drug effects on activity, and the genetic factors involved, has recently been prepared by Broadhurst. The studies we have cited are typical of many others to be found in the literature, in that many of the findings are contradictory with one another, with varying differences among genotypes (whether inbred strains or selected lines) depending on procedural details, types of apparatus, and so on. Therefore, generalizations on activity levels of specific genetically controlled animals should be treated cautiously.

An excellent example illustrating this point can be found in an article by Lassalle and Le Pape. They demonstrated that the environmental conditions associated with testing of two inbred mouse strains (C_{57}BL/6 and BALB/c) can reverse the general finding we referred to earlier that C_{57}s are more active than BALB/cs. Under seminatural and breeding cage conditions, the BALB/cs showed a higher level of activity than did the C_{57}s, and in the seminatural condition, the BALB/cs were more active in the dark).

Their concluding point is worth repeating: "A clear implication of the current study is simple and obvious, although often overlooked and neglected. Strains cannot be said to possess a given Behavioural trait. In each instance it is necessary to specify precisely the environmental conditions in which the trait was measured."

Table. Locomotor Activity: Amount and Distribution

		BAL B/c		*C57BL/6*
Seminatural environment				
	Diurnal activity	323	<	881
	Nocturnal activity	7111	>	3462
	Total activity	7325	>	4828
	Percent of total activity occurring at night	94	>	75
Breeding cages				
	Diurnal activity	1006	>	589
	Nocturnal activity	2732	>	2153
	Total activity	3734	>	2662
	Percent of total activity occurring at night	72.5	=	78

The values are medians of the different groups. Where indicated, differences are significant at $p = 0.039$ (Fisher test).

In an issue related to the genetic determination of activity, several studies have attempted to determine heritabilities and degrees of genetic determination. However, as these findings are limited to the specific genetic population measured within a specific type of apparatus and a specific set of environmental conditions, these data are of rather limited use in the generalizable information they provide for an understanding of the genetic basis of activity.

GENETIC MODELS

There are various avenues of research that have suggested that a polygenic model can best account for activity. One form of evidence for this type of genetic determination can be found in score distributions; typically, a continuous, unimodal distribution of scores has been found for most types of activity measurement when genetically heterogeneous animals were tested.

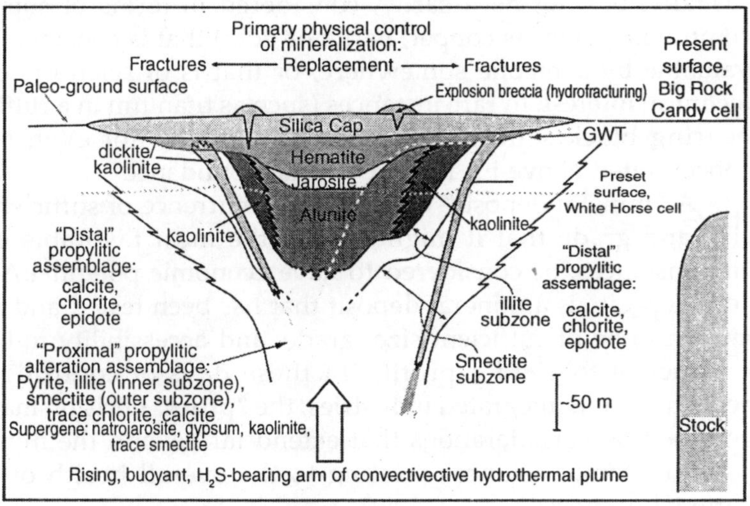

Fig. Genetic Models

Tryon suggests a polygenic model on the basis of this form of evidence alone. An excellent example of a more thorough analysis supporting a polygenic model can be found in the series of studies reported by DeFries and Hegmann. They tested large numbers of two inbred strains of mice (C_{57}BL/6J and BALB/cJ), as well as their reciprocal F_1 and F_2 hybrids and backcrosses.

The mice were tested for 3 minutes on each of 2 successive days in an automated, white Plexiglas open field, 90 cm^2. The means and variances of the transformed scores. The degree of increase in variance from the F_1 to the F_2 generation is inversely related to the number of loci affecting a trait. The variance between the F_1 and F_2 generations is very slight, especially when compared with the difference in variances between the parental strains and the F_1 generation. (One can discount the variance differences between the reciprocal F_2 s as a maternal effect, especially as it diminishes in later generations.) DeFries and Hegmann conclude that activity is controlled by genes at several loci, suggesting *at least* 3.2 loci from an earlier analysis by Hegmann.

A "mineral occurrence" is a concentration of a mineral (usually, but not necessarily, considered in terms of some commodity, such as copper, barite or gold) that is considered valuable by someone somewhere, or that is of scientific or technical interest. In rare instances (such as titanium in a rutile bearing black sand), the commodity might not even be concentrated above its average crustal abundance.

A "mineral deposit" is a mineral occurrence of sufficient size and grade that it might, under the most favorable of circumstances, be considered to have economic potential.An "ore deposit" is a mineral deposit that has been tested and is known to be of sufficient size, grade, and accessibility to be producible to yield a profit. (In these days of controlled economies and integrated industries, the "profit" decision may be based on considerations that extend far beyond the mine itself, in some instances relating to the overall health of a national economy.)

On one hand, the field observations usually begin with "mineral occurrences" (or with clues to their existence) and progress with further study to "mineral deposits" and only rarely to "ore deposits," but we must present information that helps us deal with all classes of "mineral occurrences," not just "ore deposits." On the other hand, in terms of accessible information our sample is strongly biased toward "ore deposits," for it is only in them that sufficient exposure is available to develop a real knowledge of the overall character of the mineralization process.

Some mineral occurrences are, therefore, unrecognized mineral deposits, while others are simply mineralized localities where ore forming processes were so weak or incomplete that a deposit was not formed. Thus we summarize the state of knowledge regarding ore deposit models, and we call them "mineral deposit models" with the hope that what we have learned about large and high grade metal concentrations will help us sort out all mineral occurrences to identify their true character and, we hope, to recognize which have potential to constitute ore deposits.

The attributes or properties of a mineral occurrence are, of course, those features exhibited by the occurrence. When applied to a model, these terms refer to those features possessed by the class of deposits represented by the model. It is useful to consider attributes on at least two scales: the first deals with local features that may be observed directly in the field (mineralogy, zonal patterns, local chemical haloes, and so on); the second is those features concerning the regional geologic setting and which must be interpreted from the local studies or may be inferred from global tectonic considerations (for instance, that the rock sequence under study represents a deep water, back arc rift environment, or that the area is underlain by anomalously radioactive high silica rhyolite and granite). Two of the most prominent attributes, the commodities/geochemical patterns and the mineralogy, are cross indexed to model types.

To the greatest extent possible, models were constructed so as to be independent of site specific attributes and therefore contain only those features which are transferable from one deposit to another. This goal is difficult to attain, because we do not always know which features are site specific. The term "model" in an earth science context elicits a wide variety of mental images, ranging from the physical duplication of the form of a subject, as in a scale model of the workings of a mine, to a unifying concept that explains or describes a complex phenomenon.

In this context we shall apply only the latter usage. Therefore, let us propose a working definition of "model" in the context of mineral deposits, the overriding purpose being to communicate information that helps mankind find and evaluate mineral deposits. A mineral deposit model is the systematically arranged information describing the essential attributes (properties) of a class of mineral deposits. The model may be empirical (descriptive), in which instance the various attributes are recognized as essential even though their relationships are unknown; or it may be theoretical (genetic), in which instance the attributes are interrelated through some fundamental concept.

One factor favoring the genetic model over the simply descriptive is the sheer volume of descriptive information needed to represent the many features of complex deposits. If all such information were to be included, the number of models would escalate until it approached the total number of individual deposits considered.

Thus we should no longer have models, but simply descriptions of individual deposits. Therefore, the compilers must use whatever sophisticated or rudimentary genetic concepts are at their disposal to distinguish the critical from the incidental attributes. It is commonly necessary to carry some possibly superficial attributes in order not to preclude some permissible but not necessarily favored, multiple working concepts.

One of the commonly accepted attributes of the model for the carbonate hosted lead zinc deposits of the Mississippi Valley type is the presence of secondary dolomite. But do we know that this is essential? Suppose a deposit were found in limestone; would we reject its assignment to the Mississippi Valley class? Or could it be correct that the critical property is permeability and that the formation of dolomite either (1) enhances permeability (and thereby makes the ground more favorable), or (2) reflects pre existing permeability that is exploited by both the dolomite and the ore?

Perhaps the dolomite merely records a particular range of Ca/Mg ratio in the fluid which in turn is characteristic of the basinal brines that constitute the ore fluid. In any event, the dolomite is a powerful ore guide and belongs somewhere in the "final model."

CLASSIFICATION OF MODELS USED IN THIS COMPILATION

For the purpose at hand the classification scheme has two requirements:

- It must be open so that new types of deposits can be added in the future, and
- The user must be able to find easily the appropriate models to apply to the rock and tectonic environments being investigated.

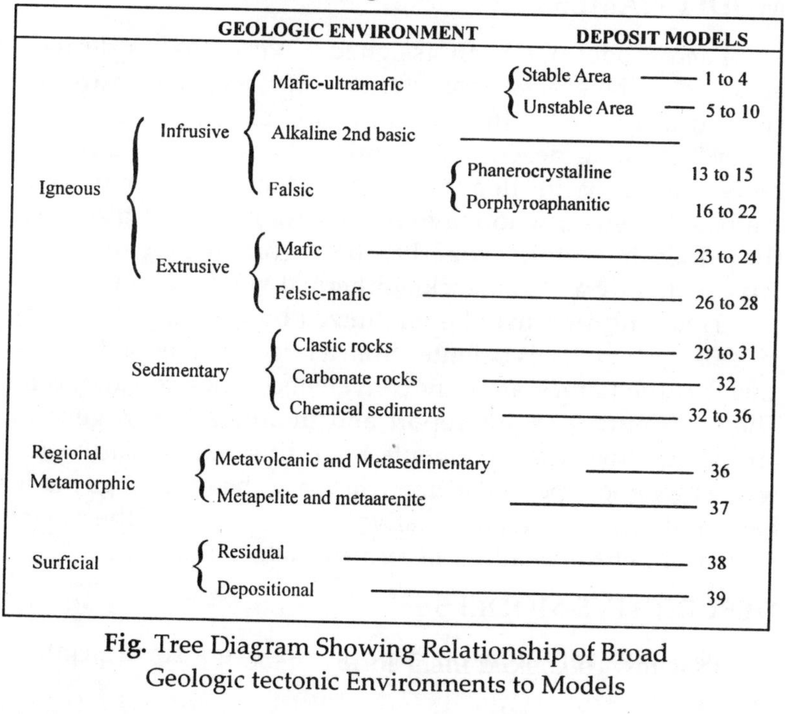

Fig. Tree Diagram Showing Relationship of Broad Geologic tectonic Environments to Models

The four logic trees that constitute a broad lithotectonic classification; this system is similar to one developed by Page and others. The classification of deposits by the environment of formation of their host rocks is continued on a finer scale. This classification scheme is relatively straightforward for deposits formed essentially contemporaneously with their host rock. However, for epigenetic deposits a conflict arises between the lithotectonic environment of the formation of the host and the lithotectonic environment of the mineralization process.

Therefore, for epigenetic deposits we have selected the most important aspect of the lithotectonic alternatives and classified the deposit accordingly. This procedure inevitably introduces a substantial bias on the part of the classifier, thus we have followed a system of including, parenthetically, alternative classifications less favored by the compiler at the appropriate alternative points in the classification scheme.

MODEL NAMES

Each model has been assigned a name that is derived either from the special characteristics of the classes or from a type locality. The latter strategy was employed to avoid excessively long descriptive names. The use of type names derived from specific deposits does produce confusion in some readers, however, who may feel, for example, that a deposit that does not look "exactly" like Comstock cannot be represented by a "Comstock epithermal vein" model.

This confusion may be minimized by realizing that most models are blends of attributes from a large number of deposits and that the names are only conveniences, not constrictions. The contributors to this report and the literature in general are not without disagreements regarding nomenclature (as well as genetic aspects and some facets of the groupings made here), but provision for alternative names is made in the model format under the heading of approximate synonyms.

DESCRIPTIVE MODELS

Because every mineral deposit, like every fingerprint, is

different from every other in some finite way, models have to progress beyond the purely descriptive in order to represent more than single deposits.

Deposits sharing a relatively wide variety and large number of attributes come to be characterized as a "type," and a model representing that type can evolve. Generally accepted genetic interpretations play a significant role in establishing model classes. Here we shall emphasize the more descriptive aspects of the deposits because our goal is to provide a basis for interpreting geologic observations rather than to provide interpretations in search of examples. The attributes listed are intended to be guides for resource assessment and for exploration, both in the planning stage and in the interpretation of findings.

The descriptive models have two parts. The first, the "Geological Environment," describes the environments in which the deposits are found; the second gives the identifying characteristics of the deposits. The headings "Rock Types" and "Textures" cover the favorable host rocks of deposits as well as source rocks believed to be responsible for hydrothermal fluids which may have introduced epigenetic deposits. "Age" refers to the age of the event responsible for the formation of the deposit.

"Tectonic Setting" is concerned with major features or provinces (perhaps those that might be portrayed only at 1:1,000,000 or smaller scale), not ore control by structures that are local and often site specific. "Associated Deposits" are listed as deposits whose presence might indicate suitable conditions for additional deposits of the type portrayed by the model.

The second part of the model, the "Deposit Description," provides the identifying characteristics of the deposits themselves, particularly emphasizing aspects by which the deposits might be recognized through their geochemical and geophysical anomalies.

In most cases the descriptions also contain data useful in project planning for mineral assessment or exploration; this aspect is especially important where limited financial and manpower resources must be allocated to the more significant tasks.

GRADE TONNAGE MODELS

Estimated premining tonnages and grades from over 3,900 well explored, well characterized deposits were used to construct 60 grade tonnage models. Where several different estimates were available for a deposit, the estimated tonnages associated with the lowest cutoff grades were used. Grades not available (always for by products) were treated as zero. Except for a few instances, the data base is so large as to preclude specific references. Several compilations of data were particularly useful sources for multiple deposit. The grade tonnage models are presented in graphical format to make it easy to compare deposit types and to display the data.

All plots show either grade or tonnage on the horizontal axis, while the vertical axis is always the cumulative proportion of deposits. Plots of the same commodity or tonnages are presented on the same scale; a logarithmic scale is used for tonnage and most grades. Each dot represents an individual deposit (or, rarely, a district), cumulated in ascending grade or tonnage.

Where a large number of deposits is plotted, individual digits represent the number of deposits. Smoothed curves are plotted through arrays of points, and intercepts for the 90th, 50th, and 10th percentiles are constructed. For tonnages and most grades, the smoothed curves represent percentiles of a lognormal distribution that has the same mean and standard deviation as the observed data; exceptions are plots where only a small percentage of deposits had reported grades and grade plots that are presented on an arithmetic scale, such as iron or manganese, for which the smoothed curve was fit by eye.

The number of deposits in each type is indicated at the upper right of each diagram. The deposits used to construct each model are listed with the model and cross indexed to model types. Correlations among grades and between tonnage and each grade are indicated only when significant at the 1 per cent level.There are important limitations inherent in the data base used for all grade tonnage models. Estimates of cutoff grades within individual deposit types can vary because of regional, national, or operator differences.

All too commonly there is no mention of the actual cutoff grades or mining widths that are incorporated; nevertheless, the grade tonnage figures given do represent material that the company or the government believed might someday be economic to mine. Stratiform deposits of large areal extent, such as phosphate or sedimentary manganese, are special problems because of differences in opinion and practice regarding how closely drilled they must be to "prove" ore tonnages and regarding the thicknesses and depths of what may be considered for eventual mining.

Effects of another source of variation, mining methods, are recognized in some of the placer models; typically, however, mining methods are fairly consistent within a deposit type. In a few instances, irregular cumulative frequency plots reflect mixing of economic and scientific data sources, such as in the plot of gold in porphyry copper deposits. In spite of the current difficulty of quantifying variation of grades and tonnages with respect to changes in cutoff grades or mining methods, the models presented here are believed to account for the main source of variation in grades and tonnages of mineral deposits variation due to differences among types of deposits.

The question of whether one counts deposits within a cluster of related deposits as individuals or as a total will probably never be resolved to everyone's satisfaction. Some geostatisticians would separate each ore body (and then argue about whether two operations on the same body should be counted separately), whereas some economic geologists would lump everything from a single district (and then argue about district boundaries). For the most part the entities summarized are individual deposits, but in some instances such data are mixed with data representing entire districts. Because of these inconsistencies, some care is necessary in comparing grade tonnage models between deposit types or in comparing this summary with those prepared using alternative methods.

Care is also warranted in interpreting the grade distributions for which data are missing; this concerns principally by product grades. In some instances, such as the platinum group element (PGE) contents in podiform chromite and the cobalt content of laterites, the fragmentary information given probably represents the entire class.

In other instances, such as the lead content of Cyprus massive sulfide deposits, the missing grades probably represent values below the lowest reported grades. The grades derived from studies of trace elements in ores more probably represent the former situation rather than the latter. Deposits strongly suspected to be small or very low grade are seldom sampled well enough to be characterized in terms of grade and tonnage, thus the sample of many deposit classes is truncated by economics.

Nonetheless, probably 40 per cent of the deposits used in these models are, in fact, non economic today; and a perusal of the figures will discover examples of both small deposits and low grade deposits.

Potential metal supply is dominated by the very few largest tonnage deposits, as shown by Singer and DeYoung (1980), who also pointed out that inverse correlations between grade and tonnage are surprisingly rare. Thus the fact that a deposit is large does not necessarily mean that it will prove to be of low grade. This means that most low grade deposits are not likely to have huge resources and also that the omission of a few low grade or small tonnage deposits will not seriously degrade the predictions of potential national supplies for most commodities.

In contrast, the missing low grade and small deposits suggest that the grade tonnage models represent a biased sample of the large number of low grade or small tonnage occurrences and prospects found by exploration. This fact must be considered in cases where the number of undiscovered deposits is estimated. In order for the estimated number of deposits to be consistent with a grade tonnage model, approximately half of the deposits estimated should have greater than the model's median tonnage or grade. Thus the probability that an untested prospect represents a significant deposit can too easily be overestimated.

OTHER TYPES OF MODELS

The bulk of this report deals with descriptive mineral deposit models and their grade tonnage counterparts, but there are other useful aspects which we wish to discuss even though we have not yet had the opportunity to develop or exploit them. They are the genetic, occurrence probability, and quantitative process models.

Many authors prefer to keep a clear distinction between descriptive and genetic models, apparently feeling that the descriptive models somehow represent "pure truth" whereas the genetic constitute a less objective philosophical position (or at least make the investigator "skate on thin ice"). It is altogether desirable to avoid confusing interpretation with fact; but it is well to remember, for example, that each time a field geoscientist extrapolates geology across a covered area he or she adds an element of "interpretation" to a "factual" map, and that this interpretation is not necessarily any more "real" (or "unreal") than, for example, an isotope geologist's conclusion that a given oxygen and hydrogen isotopic signature extracted from fluid inclusions points to a meteoric origin for the fluid.

The point is that the whole of our professional knowledge rests on a broad continuum of interpretations; many of them are so commonly accepted that they are no longer questioned, but many others still evoke challenges. Thus we suggest that a combination descriptive genetic model is not inconsistent with professional practice. The model begins as a description, but various aspects of the model become genetic as they acquire satisfactory genetic explanations. Eventually much of the model becomes genetic, as has happened, for example, with the Cyprus type massive sulfide deposits or the sandstone uranium deposits of the Colorado Plateau.

As the attributes of a model become understood in a genetic sense, the descriptive model evolves to a genetic model:

- F models are compilations of the properties of a

group of related deposits in which the reasons for certain attributes being favorable are identified. Descriptive models evolve into genetic models, and as such they become far more flexible and powerful. We have presented the three model subtypes above as if they constituted a linear logical sequence leading toward the "final" model, but in fact there must be an iterative relationship among descriptive, genetic, and grade/tonnage models. The consequence of examining any of these three may be a reassessment of the groupings of deposits chosen to be represented by a model type and the redesignation of the attributes diagnostic for that type.

With a dominantly genetic model in hand, two more model types can be generated:

- Occurrence probability models are models that predict the probability of a deposit (of a size and grade indicated by the appropriate grade tonnage models) occurring within a given area. As with the descriptive and genetic models, probability models that are tied to lithic or structural geologic entities (that is, they are genetic) are far more focused; in fact, it is probably impossible to generate a useful probability model before the establishment of a genetic model. Accurate probability models are very difficult to construct because although the technical community has very complete data on mineral producers (mines), the data on non producing mineral deposits (prospects and mineral shows) are much less well documented, a point also covered in the discussion of grade tonnage models. Even more importantly, data on barren areas are sparse.

We must extrapolate from a very fragmentary base toward a completely unseen target.

There is much to learn before the probability model can be made a dependable tool; yet the successful targeting of exploration programs by industry demonstrates that, at least on a qualitative basis, areas with better than average probabilities can be identified. It is worth noting, also, that mineral fuels are much more predictable and now can have realistic probability of occurrence values attached to specific volumes of sediments provided that the initial character and post depositional histories of the sediments are well known. It is a distant but not unreasonable dream to anticipate that some day we shall approach that level of certainty for some types of nonfuel mineral deposits.

- Quantitative process models are models that describe quantitatively some process related to mineral deposit formation; they are offshoots of the genetic model. Examples would be models of heat or fluid flow around a cooling pluton; rates of crystal growth as functions of supersaturation, impurities, and temperature; or sequences and amounts of minerals deposited from evaporating seawater.

All five of these model subtypes can be parts of the "final" model, and recycling of the model back to the original groupings stage helps refine the selection process. The persons engaged in research guidance and especially exploration and development have broad ranging needs, whereas those dealing with the availability of minerals or of land use allocation have less use for genetic or quantitative process models. Overall there is a need for a comprehensive array of mineral deposit models to meet these individual

MATURITY OF DESCRIPTIVE GENETIC MODELS

The rate at which we gain understanding and the current levels of genetic knowledge vary considerably from one deposit type to another. Such types as placers and evaporites are well known genetically and the problems in their exploration and utilization concern local site specific geologic issues rather than mineral genesis or the degree of maturation of the model.

In contrast, others such as the Coeur d'Alene Ag-Pb-Zn veins, or the massive Zn-Mn-Fe oxide/silicate bodies at

Franklin and Sterling Hill, or the Cu-U-Au at Olympic Dam, or the Cu-Zn-Pb-Ge ores of Kipushi and the Tsumeb pipe remain genetic enigmas despite, in the instances of the first two, extensive research spanning many years.

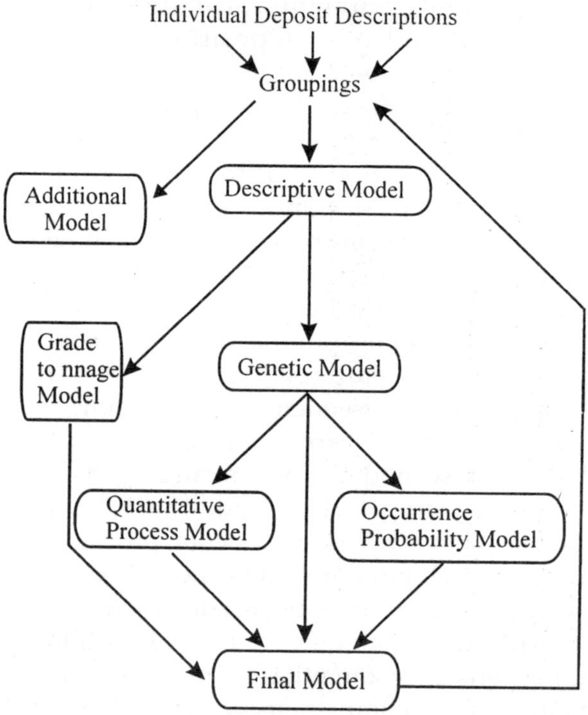

Fig. Flow Sheet Showing Evolution of Model Types. Individual model subtypes are discussed in text. It is essential that such a structure represents the repetitive cycling of information leading to continual refinement of groupings of deposits that represent each model type

Still others, such as the diamond bearing kimberlite pipes, are geologically well understood regarding their origin yet very poorly understood in terms of the reasons for their existing at any particular site. Our rate of acquisition of information is very irregular.

The several scarps between plateaus in the knowledge curve for the marine phosphate model might mark, successively, the recognition that the phosphate was a chemical precipitate, that it occurred on continental shelves where upwelling of deep marine waters occurred, and that the upwelling regions were related to wind and current patterns that were tied to the global configuration of the continents and ocean basins.

A second example from the Mississippi Valley type ores might involve scarps marking the recognition (from fluid inclusion evidence) that the ores were deposited from warm (about 100°C) highly saline solutions that could represent neither simple surface nor marine waters. A second scarp might be associated with the recognition that the deposits were integral parts of a regional hydrologic regime whose distribution and character was susceptible to interpretation.

Bring out another point: some aspects of any model always remain to be determined, thus we never acquire a "complete" model. Indeed, the approach to "complete" understanding is asymptotic, and a lot of additional effort to clear up the "last" uncertainty in a nearly perfect model is probably unwarranted.

But, new ideas and new technologies can provide the impetus for new spurts in knowledge for heretofore incomplete models. Note that the horizontal axis is simply "years of effort" devoted to fundamental geologic investigation. The scale certainly needs to be exponential in order to fit the intensively studied and sparsely studied deposit types, but this figure is strictly schematic, there being no source of documentation for either coordinate.

The figure also indicates that different deposit types may require different amounts of effort to achieve a similar level of genetic understanding.A hypothetical growth curve along which different types of deposits have been schematically arrayed.

Because some deposits (such as volcanogenic massive sulfides) are so much more difficult to understand than others (gold placers), the horizontal axis has been "normalized" by plotting a ratio of effort done to effort needed thereby permitting a smooth, although admittedly subjective and schematic, curve. There is no documentation to support this diagram, although the general concept meets with agreement among most contributors to this volume.

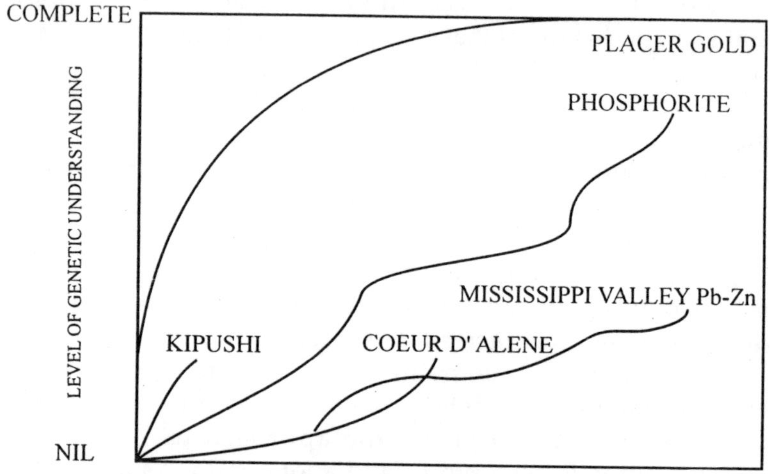

Fig. Schematic Growth Patterns for Understanding of some Typical Genetic Models

In the same paper, DeFries and Hegmann report on an activity selection study they completed. As a foundation population, they used the F_3 generation from the C_{57}BL/6J and BALB/cJ mice used in the study reported earlier. Using an open field activity measure similar to that used in their earlier study, they were able to obtain two highly active, two low active, and two control lines, with each selection (and its replication) being successful within six generations.

Of particular interest for a genetic model for activity, however, was their finding that albino rats were less active than those with any other coat coloring (agouti, cinnamon, black, and brown).

These data together with their findings that the albino BALB/c mice are less active than the pigmented C_{57}BL/6 mice, led DeFries and Hegmann to suggest that: "Although open field Behaviour is clearly influenced by genes at many loci, a major gene effect was found: albino Ss have lower activity and higher defecation scores than pigmented animals."

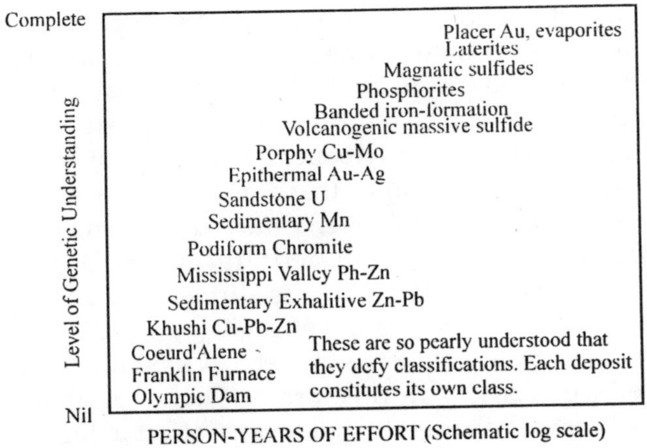

Fig. Comparison of Relative Levels of Understanding of some Important Model Types

Table: Open Field Activity Scores of C_{57}BL/6J, BALB/cJ and F_1 and F_2 Hybrid Mice

		Means [b]		*Variances*	
Generation	*Genotype* [a]	*Males*	*Females*	*Males*	*Females*
Parental	C	4.67	4.25	6.33	4.27
	B6	15.73	16.40	4.86	6.56
F_1	B6C	15.09	15.04	6.70	10.37
	CB6	13.98	13.66	15.80	21.06
F_2	CB6 x CB6	11.88	12.63	17.96	9.23
	CB6 x B6C	11.76	12.30	16.56	13.68
	B6C x CB6	12.66	12.86	15.87	16.44
	B6C x B6C	11.58	11.84	14.22	16.28

Table: Mean and Variance of Activity and Defecation Scores of Albino and Pigmented Animals in Generations $F_2 - F_5$, Pooled Across Generations

	Activity		*Defecation*		
	Mean ± S.E.	*Variance*	*Mean ± S.E.*	*Variance*	*N*
Pigmented	12.72 ± 0.18	18.65	1.78 ± 0.03	0.55	1729
Albino	10.44 ± 0.10	17.71	2.13 ± 0.02	0.49	557

This is plausible from the data presented, although it is

possible that the effect found may be due to genes linked to the c locus in certain strains, with activity not actually moderated by the c locus itself. The finding that some albino (cc) mice do not exhibit a reduction in activity warrants caution in making generalizations about the "effects" of albinism. It has been found, for example, that albino LG/J mice are highly active and that albino BALB/cBy mice do not show reduced activity when tested on an activity meter. DeFries and Hegmann used BALB/cJ mice, as did several of the other studies supporting the "albino effect."

Perhaps there is linkage of the c locus with other genes that reduces activity in this strain. There could also be a different sort of linkage with the reverse effect in highly active albino strains. At this point, however, attributing reduced activity to the lack of pigmentation seems premature.

EXPLORATORY BEHAVIOUR

PHENOTYPE PROBLEM

Exploratory Behaviour is frequently confounded with, measured by, or operationally defined as ambulation (i.e., activity), most often in an open field. Defining and then measuring exploratory Behaviour in this fashion creates obvious problems for any attempt to determine the nature of the genetic factors underlying exploration.

These problems are especially severe for the determination of specific genetic models. If one accepts the polygenic systems most often found for activity, and if one accepts the definition of exploratory Behaviour as ambulation within an open field (as many do), then the same polygenic models established as underlying activity also hold true for exploratory Behaviour. Following this rationale, the phenotypes—activity and exploratory Behaviour—are logically and empirically identical.

If activity and exploratory Behaviour have been shown to be identical—merely two terms for the same phenotype. Clearly, howevever, this is not the case. Although rather similar Behavioural measures and techniques have been used, the nature of the results obtained and especially the genetic models that these results have suggested point to the conclusion that activity and exploratory Behaviour are distinguishable phenotypes.

Even though ambulation through an open field or movements back and forth in a shuttle box are the most frequently employed measures, there is a subtle difference when these Behaviours are used to investigate exploration rather than activity. One way to separate *exploratory* activity from general activity is to measure ambulation for brief (usually 5-minute or 10-minute periods) single trials, or to count only those movements occurring when the subject is first placed in the apparatus.

The assumption here is that exploratory Behaviour is the activity that occurs in a novel environment. Another method is to measure different *kinds* of movement (e.g., rearing, as opposed to ambulation), especially when barriers or other objects are placed in the novel environment. Birke provides a good recent discussion of the aforementioned problem, and offers a similar method (using specific object exploration) for improving the measurement of exploratory activity.

The likelihood that these changes in procedure have resulted in measuring something other than "mere" activity is supported by the nature of the genetic models suggested. Although polygenic systems seem to be the rule for activity it is not uncommon to find that exploration is controlled by a single locus.

Van Abeelen

Two inbred strains of mice (SRH and SRL) were developed from two lines originally selected for high and low frequency of rearing Behaviour in a novel environment. The novel environment consisted of an open field that sometimes contained an empty food hopper. Van Abeelen created two backcross populations by crossing the high rearing frequency SRH and the low rearing frequency SRL mice, respectively, with mice from an inbred strain (DBA/2; one of the progenitor strains for the original selection study).

When he then tested the intercrosses (offspring of the two

backcross generations) in the novel enviomment, the rearing scores formed a clearly bimodal distribution, indicating the likelihood that a single locus controls this phenotype. There was also a bimodal distribution of open field activity scores, although not as striking as that obtained for rearing frequency.

This, together with the high correlations between rearing frequency (vertical activity) and open field movement (horizontal activity), led van Abeelen to conclude that the two phenotypes are both under monogenic control, and that: "The genetic factor might be allelic with the gene *Exa* [affecting "exploratory activity level"] discovered by Oliverio, Eleftheriou, and Bailey."

It does not seem farfetched to consider rearing up to investigate even a familiar stimulus (an empty food hopper) in a novel environment as a form of exploration. The argument in favour of a single locus controlling activity, and a necessary causal connection between activity and exploration, is considerably less compelling. First, mice ambulating in a novel enviomment are not necessarily being active for the same reasons as those ambulating through a more familiar environment.

Second, if mice of two genotypes differ to the extent to which they will rear up to investigate an object, they would also be apt to differ in the horizontal activity needed to approach these objects initially. (We assume they do not walk on their hind legs.) Thus, the "causal relationship" between exploration (rearing) and activity (open field ambulation) could well be an artifact of the experimental setting. Nonetheless, van Abeelen's studies do provide a distinctive approach to the genetic analysis of one type of exploratory Behaviour.

Oliverio, Eleftherious and Bailey

Using a completely different technique for genetic analysis, and also a different Behavioural testing procedure, Oliverio et al. also reported that what they termed "exploratory activity" is controlled by a single locus. The genetic technique they used involved the use of recombinant inbred (RI) strains and congenic histocompatibility lines of mice to establish a precise genetic model. In the Oliverio mice of the seven CXB (Bailey) RI strains, together with their $C_{57}BL/6By$ (B6) and BALB/cBy (C) progenitors and reciprocal F_1 hybrids were tested in a small shuttle box for 10 minutes. Shortly before testing, each subject received an intraperitoneal injection of saline. The scores on "short term exploratory activity" were measured by crossings from one end of the shuttle box to the other.

The results yielded a strain distribution pattern, confirmed by the testing of a congenic histocompatibility line, showing that short term exploratory activity is controlled by a gene linked to the *H*-26 locus on chromosome 4 (Linkage Group VIII). In this same study, they found that exploratory activity is changed (for several strains, reversed) by the effects of increasing dosages of scopolamine.

A single gene controls the changes in exploratory activity due to the administration of scopolamine. Changes in exploratory activity following injection of amphetamine were found to be due to a polygenic system. In a later study, the genetic model for short term exploratory activity (following a saline injection) was replicated, and the locus for its modification by ethanol was found: *Eam,* linked to the *H-16* locus on chromosome.

These studies are important for several reasons. The 1973 study was one of the first to make use of recombinant inbred strains (and the related techniques) to establish a specific genetic model for a *Behavioural* phenotype. Unfortunately, the influence of these same studies on our understanding of exploratory Behaviour has been largely negative, due to the misapplication of the phenotype.

The 1973 paper has been widely cited, and the gene discussed in that paper was named *Exa,* which has led to the conclusion by some that the locus controlling exploratory Behaviour (or exploratory activity) has been determined. But were Oliverio et al. measuring exploration?

It seems unlikely that either exploratory Behaviour or general activity is involved to any great extent in whatever

Behaviour is controlled by *Exa*. It hardly seems credible that a small shuttle box, hardly longer than the lengths of three mice, could serve as a novel environment for as long as 10 minutes, or even a fraction of that time. Evidence against considering shuttle box activity to be a measure of exploratory Behaviour on stimulus reactivity.

Satiation to much more complex settings has been shown to occur within that time span in several species. It would also be going beyond the data to term shuttle box crossings general activity, as Oliverio et al. do, in the absence of correlated measures.

What Oliverio et al. could well have been measuring instead of exploratory activity might have been an activity response to stress or painful stimulation. Remember that shortly before testing, each subject received an intraperitoneal (IP) injection of saline.

We attempted to replicate this study in our laboratory, using the same strains of mice and the identical apparatus and procedure, but *without an IP injection*. The results we obtained were vastly different, with not only many differences in the rankings of the strains, but also a strong indication of a polygenic effect.

There is other evidence that IP injection can have major effects on Behaviour. For example, blank IP insertions can greatly mitigate the severity of audiogenic seizures in DBA/2 mice. There seems to be sufficient contrary evidence to dispute that *Exa* is directly concerned with either exploratory Behaviour or exploratory activity.

EXPLORATION OF NOVEL STIMULI

Its movement in a particular apparatus—but no mention was made as to why the animal moves about. It could be a spontaneous Behaviour as a result of physiological changes in the animal—one theory (drive reduction) includes this presumption. Perhaps there is something of "interest" to the animal that motivates it to explore its surroundings. Before discussing exploratory Behaviour studies, it is important to understand the various components that are involved.

Stimuli have several properties that can be varied to affect Behaviour: novelty, complexity, and the interaction of these two. The novelty of a stimulus is quite important in the initiation of exploratory Behaviour of that stimulus. The more novel a stimulus, the greater the likelihood that it will be approached and examined by the animal. Given a choice between two stimuli, the subject generally will choose to examine the more novel of the two.

There appears to be a bell shaped curve in the effect novelty has on exploratory Behaviour. There is a degree of intermediate novelty that produces maximum exploration; too much or too little and the animal becomes indifferent to the stimulus or spends less time examining it One characteristic of novelty is that of change; the changing of a familiar stimulus makes it more novel.

This ties in with the concept of satiation or familiarization. If the animal is familiar or satiated with the stimulus, it will normally spend less time exploring it. If, however, some aspect(s) of the stimulus is/are changed, the animal will examine it for a longer period of time. The satiation of the subject to a stimulus or environment poses a problem in the measurement of exploratory Behaviour. As the trial or observation continues, the stimuli or environment involved in the experiment becomes more familiar and less novel until it loses its novelty almost completely.

So the Behaviour recorded at the start of the trial is more likely to be indicative of exploratory Behaviour than is the Behaviour at the end of the trial. Studies by Berlyne, Montgomery, and Thompson and Solomon have shown that exploration declines with time in a trial where the subject is exposed to the same object or environment for some specified period of time. Thompson and Solomon also found that when rats were confined in a limited space (to which they were not habituated) for a period of time, exploration initially increased before it began the ultimate decline.

This could be due to an interaction between fear and exploration and makes studies using such techniques more vague in the measurement of exploratory Behaviour. Using rats as subjects, he projected pairs of pictures of animals side by side on a screen for 10 seconds, for 10 trials. One animal in the pair reappeared constantly on one side, whereas the other side had a different animal projected each time. Berlyne found that the subjects spent more of the 10 seconds fixating on the novel pictures on the varying side than on the recurring picture.

Once the organism approaches the stimulus, the complexity of the stimulus becomes a major determinant of attention and exploration by the subject. If two stimuli are both novel, the subject will generally react more to the more complex of the two. A complex stimulus contains more variety than simple stimulus, and thus provides more ways for the subject to group the various components and come up with different responses. Several components seem to comprise the attribute of complexity (all other things being equal): the number of distinguishable elements; an increasing dissimilarity between those elements; and an inverse relationship between complexity and the degree to which several elements are responded to as a unit.

That is, the perceptual organization of the stimuli by the subject has a bearing on its degree of complexity. Clearly, the novelty and the complexity of a stimulus will integrate with each other to give each stimulus a particular characteristic. Both are variables that influence exploratory Behaviour, but the methods for manipulating each are different. The modification of a previous stimulus results in novelty and implies that the organism has an expectancy about the stimulus that is not fulfilled (when it views the modified novel stimulus).

A succession of events (stimuli) containing item(s) with short term novelty will probably be more complex during the presentation than a succession that is purely repetitious. In addition, "a pattern with a high degree of synchronous complexity will probably have a high degree of relative novelty." Assessing the involvement of novelty and complexity in experiments is quite difficult, as these two variables generally are confounded.

Dember, Earl, and Paradise offer a nice illustration of both variables. One of the pathways had its walls lined in black and white horizontal stripes, and the other walls of the second pathway were either homogeneously black or white. Sixteen rats were run for 60 minutes per day for 5 consecutive days. They found that on the first day, eight rats preferred the more complex pathway (the one with the horizontal stripes).

On Day 2, four more rats preferred the complex pathway, and the last four rats shifted preferences on the third day. None of the subjects regressed to a preference for the plain pathway. These results indicate that "any change in preference will be from the less to the more complex path.

It can be seen that even though novelty and complexity are operationally different, they can be considered to be similar in their effects. The environment in which an organism responds is as much a complex of stimuli as are the stimuli to which it is presumably responding. If the environment is unfamiliar, the organism may be responding to its novelty rather than the properties of the stimulus.

They introduced rats into a T-maze with the goal arms blocked by glass partitions at the choice point. The rats were satiated over varying exposure times and the goal arms were either both black or both white. On the test trial the colour of one of the arms was changed and the glass partition removed, allowing the subject to choose between a black or white alley. The results were as expected; the subjects chose the novel stimulus (the changed arm), indicating a "curiosity motive"as Kivy et al. stated it.

Or it could be that the animals chose the changed arm as a result of being satiated by the other arm. Dember gave support to the novelty of environment hypothesis by reversing the satiation and the test trials of the Kivy et al. experiment. What this points out is that a change in the environment, such as brightness in this case, can result in a different response by the subject.

Dember's experiment predicted and found the same Behaviours experiment, but provided a different interpretation to those findings. This shows that there can be a novel environment or novel stimuli in a familiar environment. Granted, the brightness change in the aforementioned experiment is a stimulus as well as part of the environment.

RESPONSIVENESS TO NOVEL AND COMPLEX STIMULI

Exploratory Behaviour can be viewed from several different perspectives: approach situations (exploration) and avoidance. Avoidance to novel stimuli generally taps into fear responses and degree of conflict. In these cases, the exploratory Behaviour exhibited by the subject is such that the animal avoids a particular stimulus or environment.

Normally, these studies are confounded, as it is difficult to tell if a subject is not moving because of fear or because of lack of interest. This situation is not as commonly used as approach measurement, but has been used in several studies. Regarding exploratory Behaviour as an approach situation is more common.

In most exploratory Behavioural studies, the animal can either approach a novel stimulus or not, and the resulting Behaviour is taken to be an indication of the degree of novelty and complexity of the stimulus.

Genetic Differences in Exploratory Behaviour

As in activity, genetic differences have been found in exploratory Behaviour. Harrington has done many studies on various measures of exploratory Behaviour in the rat. In runway learning, he has shown that MNRA, INR, TS3, and WAG strains were the fastest. The F344 had very poor runway performance. He also found strain differences in rats in the exploration of different environments, with MNR, ACI, and MNR-a lines having the highest latencies of emergence from a home cage into a novel environment.

Harrington brings up the dilemma of confusing a "novel environment" with novel stimuli within a familiar environment and admits that this problem still exists in the literature.

McClearn, also using six inbred strains of mice employed

four measures of activity: arena, hole in wall, open field, and barrier, to determine the generality of genetic differences. Again, both strains of C_{57}s were found to be more active than BALB/c mice on all but the hole in wall measures. In 1972, Simmel and Walker demonstrated that exploration of novel stimuli can be differentially enhanced in different strains by the presence of a companion.

We have shown that the investigation of exploratory Behaviour as responsiveness to novel/complex stimuli is certainly not new. Despite this, most Behaviour genetic studies have used activity measures for the investigation of exploratory Behaviour, with the resulting confusion regarding the appropriate phenotype.

STIMULUS REACTIVITY

A technique that permits the direct measurement of the responsiveness of an animal to novel stimuli (*stimulus reactivity*) separately from "general" locomotor activity.

Procedure. It consists of a 20-cm square arena constructed of black Plexiglas. The arena is diagonally bisected by a partition, also of black Plexiglas, with a 1-cm^2 opening in the bottom centre. This allows the subject free access between the two halves of the arena. One-half of the arena (*the novel side*) has murals consisting of black and white diagonal stripes, vertical stripes, and check erboard designs.

Fig. Arena Used for Measurement of Stimulus Reactivity and Activity

The other half (*the plain side*) has no stimuli on the walls. The entire arena is placed atop the sensors of an automated activity meter in such a fashion that identical movements on either half of the arena will yield identical responses by the meter's counter. A 10-minute trial begins when a subject is placed on the plain side of the arena. Two experimenters observe the subject by viewing a mirror angled atop the apparatus.

Operating a counter and a stopwatch, the following measures are taken: Latency (LAT), the time between the start of the trial and the subject's initial entry into the novel side; novel side time (NOV), the number of seconds (of the 600-second trial length) the subject spends on the novel side of the arena; arena crossings (ARX), the number of crossings through the hole in the centre partition between the novel and plain sides; and activity (ACT), the number of movements by the subject, as recorded on the counter of the activity meter.

Under this procedure, it is *possible* to obtain almost any combination of scores on the four measures taken. For example, a subject can be highly active on the plain side without ever venturing into the novel side; or can spend nearly all its time, inactively, on the novel side; or dart quickly back and forth between the two sides.

Multivariate Analysis of Response Measures

The first study using this procedure determined the relationship among the several dependent measures. A total of 66 male mice, 50-70 days old, were tested. The mice represented 11 different genotypes. All subjects were tested using the procedure described and, in addition, immediately after this testing, were placed in a shuttle box similar to that used by Oliverio et al. for 10 minutes.

The four arena measures (LAT, NOV, ARX, and ACT) and shuttle-box crossings (SHX) were intercorrelated and subjected to a principal components factor analysis. It can be seen that two factors emerged.

Two of the arena measures, latency and novel-side time, had high loadings on Factor I (stimulus reactivity) and loadings of nearly zero on Factor II (undifferentiated activity). One arena measure, gross motor activity, had a high loading on Factor II and the lowest loading among the four arena measures on Factor I.

Arena crossings had relatively high loadings on both factors. Both measures having high loadings *only* on Factor I (latency and novel-side time) are Behaviours dependent on the subjects' responsiveness to novel stimuli and are thus considered indices of stimulus reactivity. They are also relatively independent of gross motor activity. Arena crossings appears to measure *both* stimulus reactivity and activity. Notice that shuttle-box crossings appears to be independent of stimulus reactivity and to be primarily a measure of motor activity.

The factor structure found in the study just described clearly showed that stimulus reactivity (response to novel stimuli) can be distinguished from activity. Next, it was necessary to establish as precise a genetic model as possible, in order to determine the "fit" between the nature of the genetic determinants of the four measures and their loadings on the two factors obtained.

The technique referred to was developed by Bailey and involves the use of recombinant inbred (RI) strains for the determination of strain distribution patterns (SDPs). The Bailey RI strains were derived from a cross of two unrelated but highly inbred progenitor strains, BALB/cBy (C) and C_{57}BL/6By (B6). They are then maintained from the F_2 generation onward by strict inbreeding.

Table. Factor Loadings of Stimulus Reactivity and Activity Measures

Variable	*Factor I*	*Factor II*
LAT (latency to first novel entry)	-.641	-.011
NOV (novel-side time)	.506	-.122
ARX (crossings from plain to novel side)	.586	.302
ACT (gross locomotor activity)	-.253	.414
SHX (shuttle box crossings)	.206	.616

Inbreeding of the independent RI lines obtained from this cross results in genetically fixing, as full homozygosity is approached, the chance recombination of genes that occurred after the F_1 generation.

Unlike conventional crosses from inbred strains, this technique provides a continuing (and thus, replicable) supply of genotypically identical animals for the investigation of complex phenotypes, enabling one to establish genetic models through use of SDPs.

The seven Bailey RI strains are tested together with the progenitor strains and the reciprocal hybrids. If the progenitor strains differ significantly on the phenotype under investigation, it is possible to compile a strain distribution pattern denoting which of the RI strains resemble BALB/c (C-like), and which resemble C_{57}BL/6 (B_6-like).

It is then frequently possible to match the SDP obtained for the phenotype under investigation with one previously established for distinctive loci, such as histocompatibility. Matching SDPs indicate the possible identity or close linkage of the genes. Because it is possible for two SDPs to match by chance [$p = \frac{17}{2}$ (=.0078], it is advisable to confirm linkage by testing candidate congenic lines.

The congenic (histocompatibility) lines were developed independently from an initial cross of B_6 to C by a regimen of skin graft testing and backcrossing to B_6 for at least 12 generations. This procedure resulted in a battery of congenic B_6 lines, each line being distinguished from B_6 solely by a chromosomal segment that includes a C-strain allele at a distinctive histocompatibility locus.

Each of these congenic lines has been tested against the seven Bailey RI strains to determine which strains carry the B6 allele and which the C allele at a given histocompatibility locus.

Thus, matching an SDP obtained for any given phenotype with an SDP for histocompatibility indicates the likelihood of pleiotropism or linkage of controlling genes at the locus involved. Additional confirmation regarding the specification of the locus in question is derived from testing the congenic line that carries the histocompatibility gene of the matching SDP.

If the scores of the congenicline mice are C-like, this would confirm the control of the phenotype under investigation by a gene at this locus, as the mice of this congenic line are identical with B6 at all other loci.

The probability of an unlinked gene outside this chromosomal segment after 12 backcross generations is $\frac{1}{2}^{12}$ or 24×10^{-5}. Whether or not an SDP can be established, it is possible to obtain other sorts of information through the use of RI strains, progenitors, and hybrids.

For example, a continuous distribution of scores on a phenotype among the RI strains, together with a lack of a difference between the progeni tors, indicates that a polygenic system underlies that phenotype.

It is possible to determine dominance by the conventional method of backcrossing hybrids to the progenitors, which differ from each other; sex linkage can be established beginning with the analysis of a difference between the reciprocal hybrids, etc. In addition to the Bailey RIs, several sets of RI lines using various progenitors are currently being developed.

The study to determine the genetic analysis of stimulus reactivity tested 132 mice (which, as in the first study, were 60-day-old males), using the same testing procedure as was used for the earlier study, except that the shuttle box was omitted.

There were 12 mice of each genotype of the Bailey RI set. On the two measures that had been found in the first experiment to have high loadings on stimulus reactivity and low loadings on activity (latency and novel-side time), it was possible to establish a strain distribution pattern (SDP), with the same SDP being found for both measures.

When confirmed by testing mice from two congenic histocompatibility strains—$B_6.C\text{-}H\text{-}24^c$ and $B_6.C\text{-}H\text{-}1^b$—it was confirmed that both stimulus reactivity measures (LAT and NOV) are controlled by genes at chromosomal segments of $H\text{-}24^c$ and $H\text{-}1^b$ on chromosome 7 (Linkage Group I). The results of these tests on the congenic lines and progenitor strains.

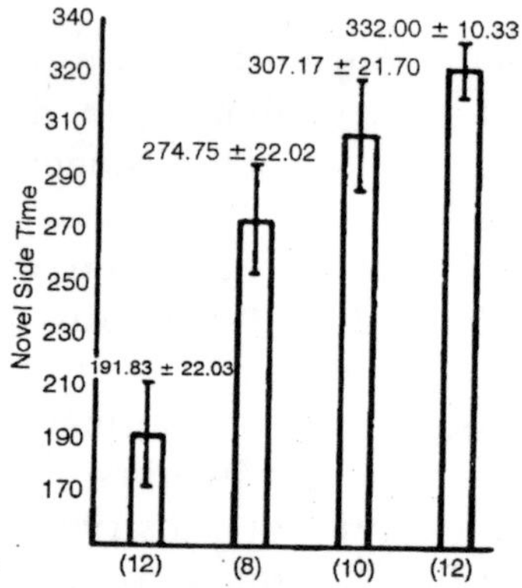

Fig. Comparison of Congenic Histocompatibility Lines with BALB/cBy and C_{57}BL/6By on (a) latency and (b) novel-side time

A polygenic model was found for the two measures with loadings on Factor II (activity and arena crossings), with the progenitor C_{57}BL/6By and BALB/cBy strains not differing from each other on either activity measure.

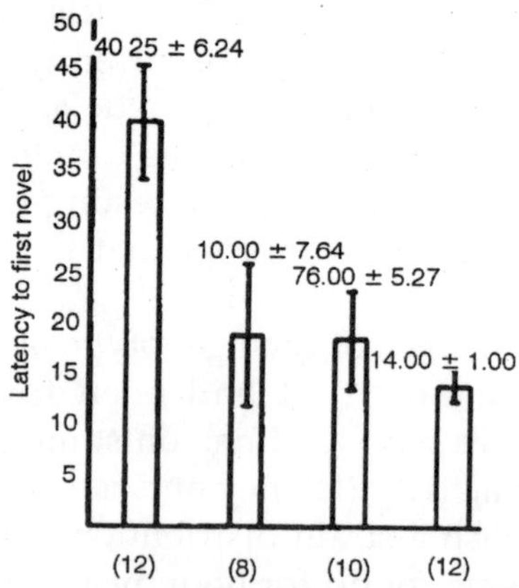

The results of the study show that the genetic model is supportive of the factor structure previously obtained: Exploratory Behaviour measured in terms of reactivity to novel stimuli can be determined separately from general activity and is lines differ from BALB/*c* but not from C_{57}BL/6, confirming the involvement of these loci on stimulus reactivity.

Generality and Stability of Stimulus Reactivity

The measures obtained in the stimulus reactivity arena appear to be quite reliable over time. The procedures and the general factor structure have been found to be basically similar when species other than mice are tested, with appropriate modifications of the apparatus. Stimulus reactivity and activity measures have been obtained for rats, gerbils, and dogs. In mice, the stimulus reactivity measures have been found to be surprisingly resistant to change due to either subject or environmental variables.

When male and female mice, either 60 or 125 days of age were tested, only genotype differences were obtained on the two stimulus reactivity measures (LAT and NOV), whereas activity differed according to age and sex. In another study C_{57}BL/6By and BALB/cBy mice were reared for several weeks in complex three-tiered cages or were placed in such complex cages one week before testing or were maintained in normal laboratory cages.

Even under these conditions of early and prior enhanced stimulation, stimulus reactivity scores (especially LAT) were largely unaffected, whereas activity scores were altered considerably.

The Importance of Exploratory Behaviour

It has been demonstrated, several decades ago as well as in the current literature, that exploratory Behaviour is a viable phenotype and can be measured as a response to novel and complex stimuli, rather than as a vague form of activity. We suggest that further research on the genetic aspects and ramifications of exploratory Behaviour would be useful, especially in two general areas.

One research area of potential importance is the relationship between exploratory Behaviour and other complex forms of Behaviour. For example, several years ago, Hayes suggested that general intelligence (in humans and in other animals) is an acquired characteristic, resulting from an individual's experience with the environment.

Individual differences between those in similar

environments are due to largely inherited differences in "experience producing drives," or the tendency to explore. Another, and perhaps more fundamental topic relates to the adaptive significance of exploratory Behaviour.

Exploration is the form of Behaviour by which an organism is most likely to obtain information about changes in its environment. Humans, rats, and mice, among other species, seem to adapt well to rapidly changing conditions in their environments, and much could be learned about the adaptive processes through further investigation of exploratory Behaviour.

Chapter 5

Behavioural Pharmacogenetics

Individual differences in both physiological and Behavioural responses to pharmacologic agents have been known for many years; the sources of these differences are often assumed to be a complex constellation of biological, environmental, experiential and for humans, sociological factors. In one sense, drug responsivity can be viewed as following a continuum; a given dose of a given drug may elicit little or no response, a moderate (often desirable) level of the same response, or an extreme response in various individuals of a patient or subject population.

These variations might be considered quantitative differences in the magnitude of a given response following drug administration. Descriptive statistics, such as ED50 (the dose of a drug that will be effective in producing the desired response in 50per cent of the subject population) and LD50 (the median lethal dose), are sometimes useful indices of the average response to a particular drug. However, the necessity of using such indices of central tendency is evidence for the inherent variability in drug responsiveness among individuals in a given subject population.

A second category of individual differences in response to drugs is evident when a proportion of individuals in a population display a qualitatively unique response to even moderate doses of the drug. This second category has often been called adverse or idiosyncratic drug reactions. An example of an adverse reaction is the hemolytic anemia produced by primaquine and related drugs in certain individuals with glucose 6 phosphate dehydrogenase deficiency. Such idiosyncratic adverse drug reactions can come about through abnormal ways in which the body acts on drugs or via abnormal states of the body that are revealed only upon drug challenge. Documentation of idiosyncratic drug reactions is prob ably greatest for drugs used in therapeutic settings, examples of which are discussed later, and more complete reviews of which are presented elsewhere.

Pharmacogenetics is an emergent discipline that seeks to elucidate the contribution of genetic factors to the types of individual differences in drug responses presented in the preceding paragraphs. The related discipline of Behavioural pharmacogenetics stems from the realization that Behavioural responses to drugs are no different in principle from physiological or biochemical responses, and further that altered biochemical processes often can be reflected in variations in observable Behaviours.

The principles and methods of Behaviour genetic analysis apply equally for evaluating differences among genotypes of mice in an open field, regardless of whether or not the mice have been previously treated with a psychoactive substance. Similarly, the methods used to attempt to describe the relative importance of genetic and nongenetic factors for individual differences in specific cognitive abilities among humans can also be applied when the phenotype in question is one related to drug responsivity (e.g., differences in responses to acute or chronic administration of alcohol).

That is, the roots of Behavioural pharmacogenetics are based in the older disciplines of psychology, genetics, and pharmacology. In its evolution, the field passed through the interdisciplinary fields of Behaviour genetics, Behavioural pharmacology, and pharmacogenetics.

In describing the nature of the then emerging discipline of psychopharmacogenetics, Fuller and Hansult and Eleftheriou referred to its hybrid nature. Fuller and Hansult went on to ask how far we might extend the metaphor by asking the relevant question: "Is this simply an area of overlap between three established areas or does it have characteristics of its own, even a touch of heterosis or hybrid vigour" Some years later, Horowitz claimed that heterosis had indeed occurred. However, he was referring to heterosis in the context of the rapid development and increasing numbers of researchers with interest in the area.

The field has begun to be realized, as evidenced by the summaries provided by Broadhurst and others. However, it has yet to be established if heterosis, in the sense that Fuller and Hansult intended, has been realized in the field of Behavioural pharmacogenetics. Does the field, in fact, have characteristics that are unique and independent of its progenitor disciplines? In the following discussions of critical issues and future directions, we hope to offer suggestions for directions that might result in this latter type of heterosis.

Until we address those issues, however, we should like to be bold enough to introduce a new metaphor to describe the progress of Behavioural pharmacogenetics. We refer to it, not only as a hybrid science, but also as a fully segregating science. The metaphorical question thus becomes whether or not the elements derived from the dihybrid cross follow Mendel's law of independent assortment. We think that they have, and that this independent assortment is evident in at least two divergent strategies that have emerged within the broader goal of Behavioural pharmacogenetics.

This broader goal is the demonstration that indi vidual differences occur in Behavioural responses to drugs, and that genetic factors must be shown to contribute to these individual differences. This primary goal has been realized by Behavioural geneticists and pharmacologists alike. As McCleam has stated, we could probably find strain differences in virtually any trait of interest to Behavioural pharmacology. The two strategies referred to differ as to how to proceed from this initial goal.

One approach is more concerned with elucidating the genetic architecture of the Behavioural response to a given drug. Its advocates suggest the application of such tools of Behaviour genetic analysis as classical genetic breeding designs, diallel crosses, and family studies, to learn more about such genetic parameters as mode of inheritance, heritability, and phenotypic and genetic correlations. The second approach uses the tools of genetics to produce (e.g., via selection) or to find (e.g., via strain surveys) populations of subjects that differ in their Behavioural responses to drugs, and then to look for underlying physiological or neurochemical mechanisms that might mediate these differences.

It should be noted that the differences between the two strategies is not so much in the initial methods employed, but rather in the interest and subsequent course of further investigation.

The former approach, here termed a biometrical approach, was the thesis for Broadhurst's recent survey of Behavioural pharmacogenetics. In this work, Broadhurst offered an admirable survey of the research to date in this field.

He further suggested that careful examinations of the genetic architecture of the traits in question have rarely and only recently been conducted. Therefore, we concentrate on the contributions made to the field by researchers employing the second approach, that of attempting to define possible physiological and neurochemical substrates of genetic differences in Behavioural responsiveness to psychoactive agents. It should become apparent, however, that the two approaches are not always separable in practice.

HUMAN PHARMACOGENETICS

Ecogenetics is a field of research defined by Omenn and Motulsky as the study of genetically determined differences in susceptibility to environmental agents. Although this broad definition includes such interesting topics as nutritional requirements and carcinogens, is the field of pharmacogenetics.

Both therapeutic and recreational drugs have effects that show considerable individual variation. A thorough review is provided by Omenn and Motulsky. We present a few examples here for the purpose of describing the kinds of pharmacological agents and genetic mechanisms that might be involved in these individual differences.

Important therapeutic drugs such as antipyrine,

phenylbutazone, and halothane are known to demonstrate marked individual differences in the attained levels in blood and in metabolism rate.

These pharmacokinetic factors are known to have a considerable basis in hereditary factors, as demonstrated by twin and family studies. Genetic control of pharmacokinetics of drugs used in psychiatric therapy is well demonstrated by the polygenic nature of inheritance of nortriptyline kinetics. Response to two classes of antidepressants, MAO inhibitors and tricyclics, is thought to reveal two genetically specific types of depression.

Responders to MAO inhibitors generally respond to all MAO inhibitors, but not necessarily to tricyclics and vice versa. The mode of inheritance was not determined in this work, but the mechanism of action is probably not one of dispositional influences, but more related to primary sites of drug action.

Omenn and Motulsky describe a dramatic condition where the widely used muscle relaxant, succinylcholine, can paralyze breathing for several hours. One in 2500 Caucasians shows this response due to an abnormal variant of plasma pseudocholinesterase, the enzyme responsible for inactivation of the drug.

This condition is nearly as dramatic as another in which a single autosomal dominant gene produces susceptibility to malignant hyperthermia induced by inhalation anesthetics such as halothane or ether, and also succinylcholine. This rapid rise in temperature often results in death.

Another question that can be addressed is the issue of a hereditary predisposition for abuse of psychoactive drugs. The answer to this question may lead us on a search for "addictable" physiologies or, perhaps, personalities. Clearly, ethanol is a drug with widely varying individual effects. Vesell, Page, and Passananti demonstrated high heritability for rate of ethanol elimination in twins (H =.98).

Yet the fact that alcoholism runs in families and probably has a genetic component is not likely to be explained on the basis of metabolism rates alone. Recent work by Schuckit, Goodwin, and Winokur not only demonstrates a reliable genetic component in alcoholism, but may be interpreted as indicating that environmental influences of sharing a household with an alcoholic are of less importance than sharing common genes. The question of etiology of this as well as other addictions may centre on the aforementioned "addictable" physiologies and/or personalities.

NONHUMAN PHARMACOGENETICS

The majority of studies using nonhuman subjects reviewed herein were designed to examine at least possible differences among groups of genetically defined subjects in Behavioural responses to various psychoactive drugs. However, it should be noted that individual differences are almost always observed, and have often been reported with interest by some investigators using random bred populations. For example, Goldstein and Sheehan found large variations among Swiss—Webster mice in levels of locomotor activity following activating doses of the opiate agonist, levorphanol.

These authors speculated that genetic factors contributed to the observed differences. Perhaps these early observations were the impetus for a selective breeding study conducted in the same laboratory several years later. Using Swiss—Webster mice as the founding population, Judson and Goldstein successfully selected for quantitative differences in changes in locomotor activity induced by an acute dose of this same opiate agonist.

Cappell and LeBlanc investigated the role of morphine pretreatment on later ingestion of a morphine adulterated solution in randomly bred Wistar rats. Individual differences in acceptance of the morphine adulterated solutions were observed in both premedicated and control groups. Those rats in the pretreatment group who were classified as "nondrinkers" could be induced to drink morphine solutions by subsequent injections of morphine and finally reached a level of morphine ingestion equal to the amount consumed by "drinkers" in the control group.

Other studies using different populations of rats have shown clear genetic differences in acceptance of morphine adulterated solutions. Although the differences in acceptance of morphine solutions observed by Cappell and LeBlanc were not established to be of genetic origin, their results are compatible with a possible genotype by treatment interaction.

Individual differences in drug responsivity have been reported also in similar populations of randomly bred mice. Ho, Loh, and Way reported significant differences in morphine induced analgesia, lethality, tolerance, and physical dependence in Swiss—Webster and ICR mice.

On several measures; there were significant differences among different populations of Swiss—Webster (or ICR) mice when these populations were purchased from different sources. Thus, although individual differences in drug responsiveness among animals of randomly bred populations are often sources of hypotheses that can be tested via genetic or pharmacologic methods, these differences may also contribute to uncertainty when attempts are made to replicate results from different laboratories.

PERSPECTIVES AND ORGANIZATION

In nonhuman subjects, genetic factors have been shown to influence responsiveness to virtually every class of psychoactive agents, including (but not limited to) stimulants, mild tranquilizers, barbiturates, alcohol, anesthetics, nicotine, opiates, and neuroleptics. Much of this work has been reviewed or collected previously by Broadhurst and Eleftheriou. Similarly, reviews of humanpsy chopharmacogenetics have also been compiled.

In keeping with the title of this volume, we have concentrated on the principles and applications of the methods of Behaviour genetic analysis with respect to their relevance to the study of individual differences in Behavioural responses to drugs. Specifically, we have concentrated on research using alcohol and opiates. It is hoped, however, that the principles, applications, and implications that we discuss will extend to the study of the Behavioural effects of any drug of choice by a given researcher.

Our decision to limit much of our discussion to the

pharmacogenetics of alcohol and opiate responsiveness goes beyond our own research interest in these two classes of drugs. First, these two classes of drugs are important from a sociomedical perspective; both alcohol and the opiates (primarily, heroin) remain major drugs of abuse. Secondly, alcohol is probably the most widely studied drug in Behavioural pharmacogenetics, and research in the pharmacogenetics of opiates is steadily growing.

Finally, there has been an increased interest in recent years in possible commonalities and interactions between alcohol and opiates. First, we present the major techniques of genetic analysis of drug responsiveness, for both humans and nonhumans. Secondly, we define phenomena of interest from a pharmacologic perspective and give examples that illustrate each from a pharmacogenetic perspective. Evaluation and interpretation are presented and discussed. Finally, we offer some personal impressions of several directions in which the field of Behavioural pharmacogenetics might be fruitfully extended by future research.

GENETIC TECHNIQUES

HUMAN STUDIES

The genetic methodologies in human pharmacology are the typical approaches in human Behaviour genetics. They may range from pedigree analysis to adoption studies and segregation analysis. We emphasize one approach that has been useful in the study of alcoholism. Detailed treatment of other methodologies may be found in standard sources. Recent work by Schuckit has emphasized a search for biological markers that might be identifiable in individuals genetically at risk for alcoholism.

Toward this end, individuals from families with a history of alcoholism respond differently on subjective reports of degree of intoxication in that they report feeling less intoxicated than controls. Physiological response to alcohol shows similar differentiation. More interesting, though, is the report that men with familial history of alcoholism exhibit higher levels of circulating acetaldehyde after ethanol consumption than controls. It is possible that some of the addictive properties of alcohol are actually attributable to its metabolite acetaldehyde.

If so, the finding of higher blood levels may be more than just a marker. It may be a prime factor in the etiology of alcoholism. At any rate, the need for such markers is obvious in order that individuals at risk for alcoholism can be closely studied. In this way those environmental factors that precipitate alcoholism may be uncovered with the use of a genetic tool.

NONHUMAN STUDIES

The major source of nonhuman subjects in Behavioural pharmacogenetic research, as is the case in Behavioural genetics in general, has been inbred strains of mice and, to a lesser extent, rats.

The process of inbreeding leads to individuals who are more closely related than are members of a population at large. In the development of inbred strains, inbreeding is secured by sequential and continuous brother—sister mating. After 20 generations of brother—sister mating, virtually all loci will be fixed for alleles in the homozygous state. Thus, all members within an inbred strain are virtually uniformly homozygous and for practical purposes, are genetically identical to each other.

However, the specific alleles that are fixed through the process of inbreeding are randomly determined (with the exception of those combinations that are eliminated by selective pressures). Therefore, although individual members within an inbred strain are (within limits) genetically identical to other members of the same strain, in all likelihood they will differ from members of a different inbred strain in the alleles fixed at many loci.

In the simplest model of quantitative genetic theory, the observable phenotype of an individual is the additive effects of its genotype and its environment. Given that two different inbred strains are reared and tested in the same (within practical limits) environment, then differences between strains on some observable trait (e.g., alcohol preference) must be determined in large part by the different alleles fixed within each. Thus, differences between two inbred strains that are significantly greater than differences within each strain (which may be due to such factors as a lack of homozygosity, microenvironmental differences, and errors of measurement) are usually construed to represent a genetic substrate for the observed phenotypic difference.

Differences between inbred strains is a useful initial demonstration for both strategies of Behavioural pharmacogenetic research. Strain comparisons have yielded significant differences in, and hence a genetic contribution to, such alcohol related responses as in preference for alcohol in a choice situation acceptance of alcohol using a forced ingestion procedure in the hepatic metabolism of ethanol and in neural sensitivity to ethanol as well as in numerous other Behavioural and biochemical phenotypes.

Similarly, genetic factors have been shown to contribute to differences among mice in preference for morphine adulterated drinking solutions in acceptance of morphine solutions using a forced ingestion procedure in readdiction susceptibility in opiate induced analgesia and hyperactivity in neurochemical responses to opiates and in the development of physical and Behavioural dependence on opiates.

For the investigator interested in further elucidating the genetic architecture of the alcohol or opiate related phenotype, these initial strain comparisons are usually extended via the tools of Behaviour genetic analysis, such as classical Mendelian breeding designs and analysis, diallele crosses and subsequent analysis, or the use of recombinant inbred strains to gain further information regarding the genetic parameters of the trait in question.

Although these analytical designs and tools have been applied infrequently in traits of pharmacogenetic interest, the reader is referred to Broadhurst for a survey of these attempts and their results.

With respect to the second general strategy of Behavioural

pharmacogenetics, the demonstration of strain differences often leads to hypotheses regarding the relationship between a trait and an underlying physiological or biochemical mechanism, or a relationship between two Behavioural traits. For example, C_{57}BL and DBA mice are known to differ in their preference for 10 per cent ethanol solutions when it is offered versus tap water in a two bottle choice situation.

These two strains also differ in the activity of hepatic enzymes of ethanol metabolism. The question now becomes whether or not these traits are physiologically or genetically related. Historically, attempts to answer such questions have relied principally on three different approaches.

The first approach might be considered one of pharmacologic intervention. For example, if the higher preference for ethanol in C_{57}BL mice is casually related to increased ability to metabolize the drug, then interfering with ethanol metabolism should reduce the alcohol preference of these mice. Schlesinger, Kakihana, and Bennett demonstrated that pretreatment with disulfuram (an aldehyde dehydrogenase inhibitor) significantly lowered the alcohol preference of C_{57}BL mice.

However, the magnitude of the decrease was slight. Similarly Ho, Tsai, and Kissin reported that C_{57}BL mice significantly incorporate higher levels of acetylcholine and greater uptake of radiolabeled choline than did DBA mice. Pretreatment with a choline acetyltransferase inhibitor, which would interfere with acetylcholine synthesis, significantly but transiently reduced the voluntary alcohol ingestion of C_{57}BL mice.

Although these two pharmacological intervention studies may contribute to our understanding of why C_{57}BLs prefer ethanol, they may tell us little of why DBA mice avoid it.

A second direction that may be taken to elucidate further a difference in two traits in two different inbred strains is to expand the number of genotypes tested. For example, Castellano and Oliverio have reported that C_{57}BL mice respond to morphine injections with increases in locomotor activity, whereas DBA mice do not.

Horowitz et al. have further shown that C_{57}BL mice prefer morphine—saccharin solutions over water, whereas DBA mice in the same situation drink almost exclusively water. Subsequently, Whitney et al. have reported a positive correlation of about. 92 between preference for morphine—saccharin solutions and morphine induced hyperactivity in six different inbred strains. This strong positive correlation was not due to differences in baseline preference for saccharin, nor to differences in level of activity following saline injections. Similarly, we have recently derived a positive relationship between ethanol preference and preference for morphine—saccharin solutions.

With respect to morphine responsivity, Brase tested four inbred strains and two randomly bred lines for hyperactivity following acute morphine administration and the degree of jumping observed following withdrawal from chronic morphine exposure. The correlation between morphine induced hyperactivity and jumping elicited by either abrupt or naloxone precipitated withdrawal was. 93-.96.

Collectively, these correlation strain surveys suggest a possible genetic relationship between the two traits of interest. However, a more stringent test would be to see if whether or not the two traits of interest are covariates in individual mice of a heterogeneous genetic population. However, many of the techniques of pharmacogenetics are invasive, if not destructive.

For example, if one were interested in the possible genetic relationship between morphine induced analgesia and the density of central opiate receptors, the use of heterogeneous mice would require measuring both traits in an individual animal.

Thus, the researcher is forced to accept the possible confounding effect of order of testing. Counterbalancing in this case is not practical, as it is as yet impossible to assay for central opiate receptors without first removing the subject's brain. Ideally, then, a researcher would like the best of both worlds: the genetic uniformity of inbred strains and the variability of a segregating population.

One population of inbred mice that comes close to meeting

these criteria consists of recombinant inbred (RI) lines. A fuller description of these mice is presented elsewhere. Basically, two inbred strains are crossed to produce an F_1 generation, which in turn are mated *inter se* to produce the F_2 generation.

Beginning with the F_2 generation, a number of brother—sister mating pairs are begun, and brother—sister mating is continued for at least 20 generations. Collectively, the surviving lines represent unique recombinations of the alleles at which the parental types were different. However, within each RI line, all members are virtually genetically identical and all alleles are fixed once again in the homozygous state.

The use of recombinant inbred lines in pharmacogenetic research has recently been reviewed by Broadhurst and by Crabbe and Belknap. The hypothetical example of the relationship between morphine induced analgesia and density of opiate receptors was actually the subject of an interesting application of recombinant inbred lines by Baran, Shuster, Eleftheriou, and Bailey. These researchers measured morphine induced analgesia and the density of opiate receptors in the brain for 11 isogenic populations of mice: C_{57}BL and BALB (the progenitor strains), the two reciprocal F_1 generations, and seven RI strains.

Their study amounted to a strain survey for each trait in question, as well as a strain ranking correlation between the two traits. Genetic differences were apparent for both morphine induced analgesia, as measured by the tail flick method, and for density of opiate receptors, as measured by stereospecific binding of radiolabeled naloxone (an opiate antagonist).

Finally, although there was a positive strain ranking correlation between binding and analgesia, the differences in opiate receptor density were not great enough to explain a significant proportion of the variation in morphine induced analgesia.

Clearly, properties of opiate receptors are one possible source of genetic differences in the ability of morphine to produce analgesia. However, other factors, including possible differences in neurotransmitter systems modulated by these receptors, are also possible contributors to the phenotypic differences in morphine-induced analgd strains.

SELECTION STUDIES

With inbred strains, the combinations of alleles that are fixed by the inbreeding procedure are random. Thus, the combinations of phenotypic expressions of these genes are also random. By selective breeding procedures, a researcher can select a phenotype of interest and selectively breed for quantitative differences in that phenotype. In research, unlike agricultural applications, it is usually desirable to select for both high and low levels of the phenotype.

Different types of selective breeding techniques are available, depending in part on the demands dictated by the desired phenotype. The basic scheme is to start with a heterogeneous founding population and to breed individuals with higher levels to others with higher levels; conversely, individuals with low phenotypic levels are bred with other low individuals. The response to selection is based, in part, on the variability in the founding population, the heritability of the trait in question, the system of selective breeding, and the selective pressure exerted by the experimenter.

Given heritable variance, the appropriate selective pressure, and enough generations of selective breeding, the procedure will produce two lines of subjects that differ markedly in the phenotype undergoing selection.

Historically, a number of selective breeding programs have been developed for various phenotypes of interest in Behavioural genetics, using both mice and rats as subjects. Many programs have selected for phenotypes initially not related to pharmacologic interests, but later evaluated for differences in response to various psychoactive agents. These studies are reviewed by Broadhurst. In addition, several selective breeding studies have been conducted in which the selection criterion was a response to a psychoactive agent; the majority of these studies used some response to alcohol as the phenotype in question.

These studies, including the Colorado sleep lines the most

and least affected rat lines the alcohol nontolerant and tolerant lines the Finnish Alko Alcohol and Nonalcohol lines and the Indiana alcohol preferring and nonpreferring rats have recently been reviewed, by workers involved in each, in a research monograph by NIAAA.

A successful selective breeding design serves at least two valuable functions. First, the demonstration that selective breeding can produce systematic quantitative changes in a drug related phenotype gives definitive proof of the contribution of genetic factors to individual differences in that particular response. More importantly, such studies often generate hypotheses regarding the selected trait and other traits thought to be related to it.

Ideally, the selection procedure should effect changes in the frequency of only those genes that are related to the response undergoing selection. Therefore, differences among selected lines in some response other than that being selected may be indicative of a physiological or genetic relationship between the two traits in question. For example, rats that have been selectively bred for susceptibility or resistance to readdiction liability to morphine also show predictable differences in their susceptibility to alcohol readdiction. These data are suggestive of a genetically based commonality in the processes underlying addiction to alcohol and opiates.

It is our contention that such demonstrations are valuable in suggesting hypotheses of commonality, but these must be tested by more rigorous experiments.

The selection index was duration of the loss of righting reflex ("sleep time") following 3.4 g/kg ethanol, although this dose was increased in subsequent generations of selection. By the fourteenth generation, there was virtually no overlap in the sleep time distribution of mice forming the long sleep and short sleep lines. Subsequent research has shown that these lines of mice differ on a number of other response measures to both alcohol and other agents.

These responses may or may not be genetically correlated to the selection index, but they furnish useful hypotheses regarding the mechanisms of action of Behavioural and physiological responses to alcohol. First, traits for which an *a priori* reason exists to suggest a relationship with the selected trait, may play no realizable role in the product of such selection.

An inverse relationship between ethanol induced sleep time and aldehyde dehydrogenase activity following alcohol challenge in the heterogeneous population from which LS and SS mice were eventually derived. This relationship would lead to the intuitive hypothesis that part of the differential response to selection in LS and SS mice might be due to differences in rates of ethanol metabolism. Yet Heston, Erwin, Anderson, and Robbins reported similar rates of ethanol metabolism in LS and SS mice.

A second phenomenon is concerned with the demonstration of concomitant responses in selected lines. LS mice have a longer duration of the loss of righting reflex following hypnotic doses of ethanol than do SS mice. On the other hand, Goldstein and Kakihana demonstrated that SS mice showed a greater severity of seizures, relative to LS mice, following withdrawal from 3 days of chronic exposure to ethanol vapors, even though blood alcohol levels were maintained at approximately the same level in both lines.

These results might be interpretable as demonstrating a relationship between acute tolerance to, and acquired dependence on ethanol. However, in the same paper, these authors further demonstrated the independence of ethanol induced narcosis and severity of seizures induced by withdrawal from chronic alcohol exposure by showing that the two traits were not covariates among individual animals of the original heterogeneous founding population from which LS and SS mice were derived.

Thus, the concomitant responses observed in the first part of their study may have been due to a stochastic association of genes subserving each response.

The final phenomenon regarding research subsequent to the development of selected lines is that, even if hypotheses derived from them are tested and and not confirmed, further hypotheses might evolve from them. Ethanol induced sleep time and severity of seizures following withdrawal from chronic ethanol exposure may not be genetically correlated characters; at least no evidence to support this hypothesis can be derieved from the results of the latter study by Goldstein and Kakihana.

However, LS and SS mice still may represent different genetic populations for the withdrawal phenotype. Horowitz and Allan investigated withdrawal from chronic mor phine exposure in LS and SS mice. Briefly, mice of both lines were implanted with sustained release morphine pellets and were withdrawn 3 days later via naloxone challenge. On two classic measures of opiate withdrawal (wet dog shakes and jumping), SS mice were more sensitive to withdrawal from chronic morphine administration.

These results were consistent with those reported for severity of seizures following chronic withdrawal from alcohol exposure, where SS mice again showed the more severe score on one index of alcohol withdrawal.

Two cautions must be taken regarding the interpretation not only of this latter study, but of all tests of concomitant responses in selected lines. First, the results may be specific to the methods and Behavioural assays employed. SS and LS mice may or may not differ with respect to their general level of physical dependence on alcohol or morphine following these drug regimens.

Secondly, as Satinder and others have pointed out, the true test of a genetic relationship between two traits is to see whether or not the traits are covariates in individual animals of a population of heterogeneous genetic stock. Thus, the relationship suggested by the results of the studies by Goldstein and Kakihana and Horowitz and Allan remain to be evaluated.

HETEROGENEOUS STOCKS

Genetically heterogeneous stocks are found in several different forms, each of which has its peculiar advantages and disadvantages. The common characteristic of these stocks is a lack of genetic uniformity from animal to animal. Within that definition, however, is a wide range of genetic diversity. The presence of this diversity provides several tools for the pharmacologist interested in the control of but not specific investigation of genetic variables. For the geneticist, these stocks have a wide variety of uses. First, we distinguish several types of stocks available.

"Random Bred" Stocks

Many commercial suppliers of laboratory animals, particularly rodents, characterize their populations as random bred. The extent of genetic variability in most of these stocks is not known. In house breeding colonies maintained by investigators are likely to be rather inbred owing to founder effects. This fact and general lack of information on ancestry of these types of stocks make them less suitable for genetic work than the other types of stocks described here. However, Rice and O'Brien have characterized the often used Swiss mouse.

This work suggests more genetic variability (enzyme polymorphisms) in this stock than is suggested by the foregoing. The usual lack of controlled breeding schemes and the many different populations of Swiss type mice call into question the generality of this finding.

Inbred Strain Crosses

The production of an F_2 generation from two inbred strains is relatively inexpensive monetarily, as well as in time, and provides a heterogeneous stock with an amount of genetic variability determined by the number of allelic differences in the two parental stocks. Maintenance of such a stock by production of further segregating generations could be accomplished by a random mating scheme with some care taken to avoid inbreeding.

Such a stock with controlled breeding would be suitable for a variety of genetic analyses. The degree of usefulness is limited by the number and character of differences between the two parental stocks. More than two alleles at a locus would not be possible. Choice of progenitor stocks should attempt to maximize genetic diversity for the phenotype(s) of interest or in general characteristics if genes underlying such phenotypes are not identified. The latter approach would simply suggest using maximally unrelated strains.

The Eight or Four Way Cross

The limitations suggested for the F_2 type stocks may be overcome by complex crosses involving more than two parental stocks. Such crosses have several times served as the foundation stock for selective breeding experiments. Their usefulness can, however, extend beyond this role. McClearn, among others, has suggested the use of the eight way cross. Beginning with eight inbred strains of diverse origins it is possible to construct a stock with an extremely high degree of genetic diversity, expressed in a variety of phenotypes and specifically identifiable in numerous enzyme polymorphisms.

It is conceivable that within such a stock, allelic combinations might arise that would not be possible in natural populations. This could occur if some of the progenitor strains were derived from completely separate populations. For example, recently inbred strains from *Mus musculus castaneus* and *M.m.molossinus* both show allelic variants at some loci that do not exist in any of the more commonly used strains.

Such uniqueness might very well provide useful tools for the analysis of drug action. This notion is analogous to the argument that single gene mutants may provide information on "normal" nonmutant physiology. In this case however, the degree of phenotypic aberration may provide either qualitative or quantitative uniqueness of response.

Experimental Advantages

For the pharmacogeneticist, the major advantage of a genetically heterogeneous stock may be the appropriateness for investigations of genetic covariance or correlations. These correlations may involve several Behaviours affected by a single pharmacological agent or single phenotype effects of several agents. Although the number of subjects required for such analyses may be large, and vary in a negative fashion with character heritabilities benefits of such work may be large. In the present context, heterogeneous stocks provide a clear advantage over the use of inbred strains in such analyses. The fact of complete homozygosity in such inbred strains produces a physiologically aberrant situation. It is conceivable that pharmacological effects might depend not only on allelic genotype but also on the fact of homozygosity. More importantly, the random fixation of allelic pairs as a result of the inbreeding process may have produced unique polyallelic combinations.

To the extent that these genotypic profiles might produce correlated Behavioural/pharmacological effects, potentially large genetic correlations could result. These relationships might however be transient effects that could disappear or diminish in a random breeding population where they would occur with a smaller frequency, as those genotypic combinations would be rare. The presence, then, of genetic correlations in heterogeneous stocks indicates that they are probably more general phenomena.

Another experimental approach that finds powerful application has been termed the genotypic method in Behaviour genetics. The identification of an allelic variant can allow assessment of the pleiotropic effects of such on Behavioural response to drugs. This approach is most often used with coisogenic stocks, but could be accomplished in heterogeneous stocks as well. For example one might identify electrophoretic variants of mouse liver alcohol dehydrogenase, which are known to be under Mendelian control and proceed to assess alcohol responsiveness in mice of differing genotype.

If the phenotype examined were known to be influenced by dispositional variables, then one might expect an influence of ADH. Such pleiotropic effects could be examined for a wide variety of identified allelic variants (i.e., coat colors or enzymes). The advantage of the heterogeneous stock in this case is the ability to examine effects of one genetic variable (the marker phenotype), while maintaining a more or less random genetic background on which it is found.

Again, observed effects might be more general than those

seen with coisogenic stocks. The utility of heterogeneous stocks for nongenetic research also deserves comment. Although the advantages of using inbred strains derives from their reproducibility and isogenicity, these variables are also potentially disadvantageous for the researcher not interested in genetic mechanisms.

Aside from the generality issue raised by pharmacological work with one or few strains, the fact of genetic homogeneity at all loci may be less than desirable. Simply stated, a completely homozygous mouse is clearly an aberrant type of animal. Inbreeding depression effects may very well extend to pharmacological responsiveness, particularly if the Behavioural phenotype is related to fitness.

The heterogeneous stock provides a physiologically more normal animal. In stocks produced by careful crossing of inbred strains (i.e., 2–, 4–, or 8–way crosses), the genetic ancestry of any particular subject can be as well known as that of any inbred animal. Contrary to popular belief, the variability associated with measurement of a phenotype is not necessarily larger and can be smaller in heterogeneous stocks than in inbreds. The good general health and large reproductive capacity of heterogeneous stocks is also a consideration.

An interesting use of the Colorado HS mice was reported by Reed. In a population of 1055 mice, heart rate was measured before and after administration of ethanol. What occurred was a "normalizing" response. Mice that had deviant baselines scores tended toward the mean on the postalcohol test. Thus a substantial negative correlation between preethanol and postethanol scores existed.

This normalizing response was also shown to exist in humans. In another study with HS mice it was suggested that effects of ethanol on body temperature, open-field activity and heart rate were not genetically correlated.

MUTANTS AND COISOGENIC STOCKS

The genotypic approach in the preceding as received relatively little attention from the experimental pharmacogeneticist. This is not for a lack of genetic material to work with. Numerous allelic variants of enzymes have been identified in a variety of species. The mouse, as a mammal, is probably the species of choice here as well.

A variety of enzymes could be examined for allelic variant effects on pharmacologic responsiveness. In addition, the presence of numerous neurological mutants provides an obvious subject matter for neuropharmacological if not Behavioural investigation.

Two examples of mutants are pertinent. First, the single-gene neurological mutant, Jimpy, is at least 10 times less sensitive to the analgesic effects of morphine. These mice also have fewer opiate receptors. Whether these effects are direct or executed during development is unknown. A second example is not an identified mutant, but a plausible one. Poley reported that an Alberta subline of the C_{57}BL/6 mouse would not consume ethanol in a choice situation the way that other closely related substrains would.

Thus, although both C_{57}BL/6J and C_{57}BL/10J mice consumed considerable ethanol, the C_{57}BL/ 6UAE mice did not. Whitney and Horowitz replicated that work, but also showed that the Alberta subline would consume as much sweetened morphine as would the other sublines. The difference in ethanol consumption is certainly suggestive of a single-gene difference.

PHARMACOLOGICAL PHENOMENA

INITIAL SENSITIVITY

Naive sensitivity to alcohol may be assessed in a number of ways depending on the Behaviour or process of interest. Loss of the righting reflex in the mouse has been a useful phenotype for genetic study. Although some variation in procedure occurs from lab to lab, the basic method is that of Kakihana, Brown, McClearn, and Tabershaw.

After injection, the mouse may or may not be assessed for the length of time needed to fall from a suspended grid (fall time). The primary measurement is the time it takes the subject to right itself spontaneously three times in 30 seconds.

This measure gives a general estimate of hypnotic potency of a depressant drug. When combined with assessment of blood levels of the drug at time of waking, an estimate of neurosensitivity is obtained. This, of course, was the criterion phenotype for the selective breeding programme for the LS and SS lines. Extensive study of strain variation in this response to alcohol also exists.

Measures of intoxication are numerous. The tilting plane test, the grid test, and the Rota-Rod have all been utilized in pharmacogenetic work. The grid test involves suspension of a hardware cloth grid over a metal plate. Placement of a paw through the grid onto the plate results in a circuit activation and can be registered as a count.

The more intoxicated a mouse, the greater number of "step-throughs." When combined with simultaneous assessment of locomotor activity, a relatively sensitive measure of intoxication can be obtained. That approach determined LS mice to be relatively more intoxicated by subhypnotic doses of ethanol than were SS mice. Dose-dependent effects may be obtained but, at times, only if a correction for general activity levels is made.

A third measure of initial sensitivity examines the purported biphasic shape of the ethanol dose-response curve. Locomotor activity is generally stimulated by low (1-2 g/kg) doses of ethanol. Higher doses of ethanol produce depression. The stimulant properties of low doses are of interest in that they seem to present a paradox: stimulant actions by a depressant drug.

The genetic studies of this Behaviour at times indicate quantitative genotypic differences and in other cases qualitative differences. For example, both LS and SS mice are activated by 1.0 g/ kg, but higher doses depress LS mice and continue to activate SS mice. Alternatively, BALB/c mice show activation to several low doses and C_{57}BL/6 mice apparently cannot be stimulated at all.

These examples were chosen to indicate that if an understanding of underlying mechanism is of interest, then it is probably best to study simple Behaviours. Study of complex

Behaviours, although important, is difficult from a genetic viewpoint because more genes, and thus more physiological systems, are probably involved.

TOXICITY

Both alcohol and opiates should be seen as toxic agents. Cirrhotic livers and heroin overdoses are *prima facie* evidence. As such, standard toxicological techniques can be applied. A study by Whitney and Whitney showed mortality rate among 10 genotypes to be negatively related to their free choice preference for ethanol. Also, the LD50 values for LS and SS mice were equivalent in mice from the 14th generation.

Another type of toxicity is studied in the conditioned taste aversion experiment where ethanol can serve as a UCS or poison to produce flavour aversions. Toward this end, Horowitz and Whitney showed DBA/2 mice to form much more severe conditioned aversions, when ethanol was the UCS and saccharin flavored water the CS, than did C_{57}BL/6 mice. Of similar interest is a recent report that BALB/c mice do not form conditioned taste aversions when doses of ethanol up to 6.0 g/kg are used as the UCS.

Chronic Drug Tolerance

This refers to the waning of responsiveness with repeated administration of a drug. It can be expressed in two ways. First, tolerance is evident if the magnitude of the response decreases with repeated administration of the same dose of the drug. Alternatively, tolerance can be demonstrated if, with repeated exposure, greater doses of the drug are required to maintain a given magnitude of the response being evaluated.

Tolerance observed at the Behavioural level may be a result of alteration of drug availability (e.g., metabolic tolerance), adjustments made by the organism to compensate for the effects of the drug (e.g., Behavioural tolerance), or changes that occur at the site of action due to continued presence of the drug (e.g., pharmacodynamic tolerance). Genetic factors at any of these levels could contribute to individual differences in the degree of tolerance to either alcohol or the opiates in both humans and nonhumans. Pharmacogenetic aspects of drug tolerance and dependence, including both alcohol and the opiates, have been reviewed extensively in an excellent recent summary article by Crabbe and Belknap. Thus, one example is presented for genetic differences in the development of tolerance to alcohol and to the opiates.

Moore and Kakihana examined the development of tolerance to the hypothermic effects of ethanol in three inbred strains of mice. Animals were injected with 3 g/kg ethanol daily for a period of 8 days. Rectal body temperatures were recorded on the first, third, and last day of drug administration. As a general summary, tolerance developed to the hypothermic effects of alcohol in mice of the BALB and C_{57}BL strains, but not in DBA mice.

In fact, DBA mice may have become more sensitive to the hypothermic effects of ethanol with repeated administration. Although interpretation of the degree of tolerance is complicated by the magnitude of hypothermia induced by the initial exposure to ethanol in mice of these three genotypes, it is clear that genetic factors affected the rate and direction of changes in body temperature resulting from repeated ethanol administration.

Although differences in alterations in alcohol elimination rates were apparent in this study, these metabolic differences were not great enough or consistent with the genetic differences in the pattern of tolerance to the hypothermic effects.

Tolerance to the analgesic effects of morphine was evaluated by Oliverio, Castellano, and Eleftheriou. Both initial sensitivity and tolerance were evaluated using the hot-plate technique in the Bailey BALB, C_{57}BL, F, and recombinant inbred lines. Mice were initially tested following three doses of morphine and then made tolerant by four additional administrations of morphine.

Tolerance evaluation followed the fifth morphine injection. In general, genetic differences were apparent in both initial sensitivity and acquired tolerance to the analgesic effects of morphine as measured by response to thermal stimulation. However, for both the initial sensitivity and the tolerance test, analgesia was evaluated at several postinjection times. Thus, interpretation of the results is complicated by dose by postinjection trial, which clearly could be influenced by the repeated-measures design.

Dependence

Dependence is a consequence of chronic drug administration in which the drug is needed in order for the organism to maintain normal functioning. Both alcohol and opiates can lead to dependence following repeated administration. As Crabbe and Belknap have pointed out, dependence is inferred by the consequences of terminating drug administration (i.e., withdrawal or the abstinence syndrome).

Genetic differences in the degree of dependence on alcohol and morphine have been alluded. Goldstein and Kakihana reported greater seizure severity following withdrawal from chronic alcohol exposure in SS than LS mice, whereas Horowitz and Allan found analogous results in mice of these lines in both jumping and wet dog shakes induced by withdrawal from chronic morphine administration. Similarly, the demonstration by Brase et al. of genotypic differences in severity of opiate withdrawal of the use of inbred strains.

Self-Administration

This is another phenomenon associated with chronic drug administration. It can be demonstrated in a number of different paradigms, but in each case, the organism determines the amount of drug to be administered. Due to the number of subjects required in pharmacogenetic studies using nonhuman subjects, the most common techniques involve some variation of oral ingestion.

When the procedure employs an alternative to consuming a drug-adulterated solution or food source, drug preference or drug seeking can be examined. Genetic differences in preference for alcohol and morphine-adulterated solutions among geneticaly defined subjects. A cautionary note should be made regarding the importance of task specificity when evaluations are made regarding the genetic contributions to differences in tolerance, dependence, and self-administration.

These cautions will be expanded in subsequent. First, degree of tolerance for a psychoactive drug is often dependent on the choice of response used as a criterion of strength of effect. Secondly, inasmuch as dependence is inferred from the consequences of drug termination, genetic differences may reflect differences in the response being measured rather than differences in the absolute degree of dependence.

Finally, oral ingestion regimens of self-administration may be confounded by factors unrelated to the neurochemical effects of the drug, such as preabsorptive orosensory cues. Although all these task-related factors may be of interest from a Behavioural pharmacogenetic perspective, such factors must be considered in making statements regarding genetic differences in the more general phenomena under investigation.

LOCUS OF EFFECT

Absorption and Distribution

When psychoactive agents such as alcohol and opiates are administered systemically and when genetic differences are observed in response to such administration, it is tempting to attribute the observed differences to genetically determined differences at the site of action, usually in the central nervous system. However, such conclusions are predicated on the assumption that equal amounts of the drug reach the site of action.

The assumption may not be warranted, if genetic factors contribute to different amounts of the drug reaching the central nervous system via such mechanisms as differential absorption, metabolism, and permeability of the blood—brain interface.

Many careful attempts to evaluate such possible differences in drug distribution have been conducted in pharmacogenetic research involving alcohol and the opiates. For example, Brase et al. measured morphine induced hyperactivity and analgesia (using an abdominal constriction technique) in four inbred and two randomly bred lines of mice.

These researchers also measured brain morphine levels at the time of each test. Strain differences were apparent for all measures, including brain morphine levels. However, brain morphine levels did not correlate significantly with either analgesia or locomotor activity, indicating that the observed differences in these latter two phenotypes could not be ex plained by differences in distribution of the drug. Similarly, Moore and Kakihana demonstrated differences in blood alcohol levels following ethanol administration in three inbred strains of mice, but further showed that these differences could not totally account for differences among these strains in ethanol induced hypothermia.

Finally, Goldstein and Kakihana equated blood alcohol levels in LS and SS mice, but still found differences in the severity of seizures following withdrawal from chronic alcohol exposures.

In a slightly different sense, issues of drug absorption may be of primary interest in the investigation of Behavioural phenotypes in pharmacogenetic research. For example, the substrate for the difference in alcohol preference between C_{57}BL and DBA mice has been the subject of considerable research. Recently, Belknap and his associates have suggested that different mechanisms may underlie the avoidance of alcohol by DBA mice, and the preference that has been consistently demonstrated by most sublines of C_{57} mice.

Specifically, DBA mice may avoid alcohol on the basis of preabsorptive (e.g., taste and olfactory) cues, rather than on the pharmacological consequences of ethanol. C_{57}BL mice, on the other hand, may be less sensitive to these preabsorptive cues, and thus postabsorptive cues may be more relevant in explaining alcohol preference in mice of this genotype. Thus, it might be of interest to see whether a specific subline of C_{57}BL mice (C_{57}BL/Uae), which have been reported to avoid 10per cent ethanol solutions, show DBA like responses to the preabsorptive cures of ethanol demonstrated by Belknap and associates.

Peripheral Effects

Although drug action on Behavioural phenotypes is often assumed to occur via effects on the central nervous system, it should be apparent from the preceding that genetic influences on these drug responses may operate through peripheral mechanisms. Dispositional and pharmacokinetic factors are, however, not the only characters of this type.

Cardiovascular as well as autonomic nervous system effects of pharmacological agents are well known. It is conceivable that alteration of such characters as heart rate, blood pressure, and sympathetic tone may play a large role in the general Behavioural response to a drug. In other words, subjective evaluation of drug response may depend on information about changes in activity of these peripheral systems.

A probable genetic component in variability of peripheral response to alcohol is well known. Racial differences in such variables as heart rate response, blood pressure effects and facial flush response are documented. The greater sensitivity of Orientals to alcohol may be due in part to metabolic factors. However, the marked differences in peripheral response to alcohol may be related to differential drinking patterns as well.

The issue may be even more complex, as suggested by Zeiner, Peredes, and Christensen, who investigated the role of the primary metabolite of ethanol, acetaldehyde, in these racial differences in peripheral response. If acetaldehyde is responsible for ethanol effects, then examination of its direct effects is required.

CENTRAL NERVOUS SYSTEM

Within the central nervous system (CNS) there are a number of physiological/ biochemical loci where genetic influences and drug effects might converge. One important way in which such genetic influences might occur. Genetic control over CNS development and the general issue of epigenetic processes. There remain several ways in which the genome can exert an influence over functional states of the CNS and, both by demonstration and inference, Behaviour.

Membranes and Channels

The unique attribute of the neuron, its excitable membrane, depends on the regulated passage of several kinds of ions through specialized protein channels. The action potential is in reality the generation of sodium and potassium currents as they traverse the neural membrane through these channels. The release of several neurotransmitters is dependent on similar actions of calcium.

Modification of the number of these channels or their willingness to pass respective ions would affect neural function at its most basic level. Genetic control of affinity might be direct, given a single amino acid change due to mutation. Channel number control would seem to depend on gene regulation or posttranscriptional/epigenetic influences.

Several channel mutations have been isolated and mapped in paramecium species. Although this is not an example of a neural phenomenon, it remains an impressive fact that calcium and potassium channel mutants have Behavioural sequelae in these single celled organisms. A particularly interesting use of these mutants was reported by Schein, who took advantage of asexual reproduction and induced autogamy to follow alterations in membrane excitability during changes of paramecia from heterozygote to homozygous mutant form.

Such a procedure allowed estimation of the calcium channel half life at 5-8 days. Although no such demonstrations are, to our knowledge, available in neural tissue, the importance of such a possibility cannot be overlooked.

It has been long known that the narcotic effectiveness of depressant agents such as alcohol is related to the degree of lipid solubility. Thus the degree to which a compound can distribute itself into neural membranes is a direct predictor of its anesthetic potency. Because it is well accepted that anesthetics and many sedative hypnotic drugs can act in this way, it is a parsimonious assumption that genetic variation in membrane construction could result in differing lipid solubility of the same drug in different genotypes.

Although such genetic variation has, to our knowledge, not yet been demonstrated, the demonstrable hereditary variation in neurosensitivity to alcohols seems likely to involve these membrane mechanisms. The direct action of anesthetics on membranes is thought to occur via alteration of the proteins found there; these proteins are likely to be those of the channels.

One investigation in this area has attempted to use genetic variation as a tool for studies of possible commonality of action of ethanol and halothane, a gaseous anesthetic. The LS and SS mice described earlier were shown to be similarly affected by halothane, even though their ethanol response is so strikingly different.

It was concluded that the two drugs act via different mechanisms. In the present context, it would appear that different classes of membrane proteins are being affected by the two drugs. Alternatively, the differential LS and SS response to ethanol may not take its origin in membrane effects.

Receptors

In recent years it has become increasingly apparent that the functioning of the brain is critically dependent on highly specialized receptor molecules, protein in nature, and thus potentially subject to genetically controlled variation. These receptors are usually thought of as specific for particular neurotransmitters, neuromodulators, or hormones. Several clear examples of the involvement of such receptors exist in the psychopharmacogenetic literature. We deal here with drug effects and receptor studies of three types: benzodiazepines, opiates, and barbiturates.

Benzodiazepines have recently been shown to bind to specific sties in CNS membrane fragments. Thus it appears that the nervous system may possess endogenous neurotransmitter systems that are similar in function to the functions affected by this class of tranquilizers. The clinical efficacy of benzodiazepines (i.e., their anxiolytic effects) tends to correlate well with their affinity for these specific receptors.

It has been suggested that characteristics of such endogenous benzodiazepine systems might be altered in pathological disorders of affect that are usually treated with this drug class. In animal model research, demonstrable genetic variation in emotionality/reactivity seems a likely candidate for explanation by variation in these receptors. Toward this end, Robertson, Martin, and Candy have studied benzodiazepine receptors in the rat strains selectively bred by Broadhurst for reactivity and nonreactivity.

The Maudsley non reactive rats (MNR) had a higher specific binding of diazepam in every brain region examined than did the Maudsley reactive (MR) rats. The largest differences were found in limbic structures.

The nature of the differences was primarily one of receptor number rather than affinity. In hypothalamic tissue the number of binding sites was 1000 fmol/mg protein in MR rats and 1385 fmol/mg protein in MNR rats. The affinity for the ligand was only slightly higher in the MNR strain: $K_D = 3.64$ nmol for MR and 3.24 nmol for MNR. These authors suggest this issue of receptor density as a plausible basis for the reactivity differences in the strains.

Presumably the more reactive MR rats have a less extensive endogenous "tranquilizing" system. In another study from the same laboratory similar results were found in four mouse strains. "Emotional" BALB/c mice had a lower density of benzodiazepine receptors than did three "less emotional" strains. Both these studies reinforce the notion that endogenous benzodiazepine systems play a role in emotional Behaviour. A more complete analysis in the future will, we hope, combine this type of study with actual examination of benzodiazepine effects on "emotionality."

Another type of receptor, the opiate receptor, has been similarly studied. Considerable genetic diversity has been demonstrated for Behavioural effects of opiate compounds in mice. Again, a parsimonious prediction might be that opiate receptor density or affinity could serve as the basis for varied

Behavioural effects of opiates. Two groups of investigators have directly addressed this issue.

Baran et al. investigated naloxone binding and analgesic response to morphine in C_{57}BL/6By, BALB/cBy, their reciprocal F_1 hybrids, and seven recombinant inbred strains derived from those parental stocks. Considerable strain differences for total stereospecific naloxone binding in whole brain and also analgesic response were reported. Even though the strain with the lowest naloxone binding (CxBK) also had the lowest analgesic response, the correlation between binding and analgesia across strains was a non significant.48.

The relatively low statistical power for this correlation suggests that a real, but minor relationship may exist. Crabbe and Belknap point out that genotypic variation in Behavioural response may well arise in part from dispositional factors, and thus the relative importance of receptor characteristics would diminish.

We would suggest two additional possibilities. First, the Baran et al. study did not examine specific brain regions. To the extent that analgesic effectiveness of morphine is mediated by specific brain regions (e.g., the periaqueductal grey), analysis of a heterogeneous whole brain could well obscure important predictability from specific regions.

The notion of nonhomogeneous opiate receptor system distributions raises a second issue. Present conceptions of opiate systems within the brain assume not only anatomical topographies, but biochemical heterogeneity of opiate receptor subtypes. It is conceivable then, in this context, that naloxone specific binding may not reflect a more specific receptor subtype, which holds importance for the genetic variation in analgesic potency. Use of another ligand might reveal a correlation with the genetic effects on analgesia.

A recent study has examined the role of subpopulations of opiate receptors in genotype influenced morphine sensitivity. This work with C_{57}BL/6 and DBA/2 strains of mice found that strain differences in opiate receptor number depended on both ligand (and thus subpopulation) and brain region. The C_{57}BL strain had considerably more striatal opiate receptors when D-Ala_2 -Met-enkephalin or Leu-enkephalin were used as ligands. Such differences did not occur when naloxone or di hydromorphine were ligands, nor did they occur in three other regions examined.

These results suggested that the locomotor response to morphine in C_{57}BL mice could take its origin in this striatal subpopulation of opiate receptors. It is interesting that neither of these studies on strain differences in opiate receptor characteristics reported differences among genotypes in affinity (K_D) of the ligand for the receptor.

This suggests that the kind of genetic variation present may not be that of structural gene mutation but rather complex regulation of the number of receptors inserted into neural membranes, or perhaps cell number. In this context, it may well be that control of receptor number in neural tissue could serve as a prime research area for regulation of gene action in higher organisms. A developmental perspective on this issue would probably also be of value.

As a third example of receptor characteristics and genetic control of drug response, the work of Waddingham, Riffee, Belknap, and Sheppard is of interest. The regulation of membrane bound â-adrenergic receptors by catecholamines and the ubiquitous nature of norepinephrine influences on a variety of Behaviours suggest these receptors to be of import in neural function.

This study examined genetic influence on these â receptors and their response to phenobarbital treatment. C_{57}BL/6 mice had generally higher densities of â adrenergic receptors in several brain regions. After phenobarbital treatment, receptor proliferation was apparent, but dependent both qualitatively and quantitatively on strain and brain region. Again here, regulatory gene activity may be of interest particularly in the framework of an inducible system—induction produced by phenobarbital treatment.

Enzymes

As major gene products, enzymes in the CNS are likely candidates for heritable variation. The range of enzymes active in neural tissue is wide and provides a wealth of material for pharmacological study. At a time when research on Behavioural control by various neurotransmitter systems is so predominant, it is perhaps easy to overlook other kinds of enzyme systems that could markedly influence drug response. Not the least of these are those biochemical pathways involved in energy metabolism.

Small perturbations in cerebral energy metabolism can have potentially major effects on a wide variety of Behaviours. The example provided by hypoglycemia and its effects on cognitive and affective characters is clear. The brain is a high energy utilization organ. All specialized functions of neural tissue depend on the more basic biochemistry of energy sources. Although we are not aware of genetic studies of enzymes involved in such energy metabolism and pharmacological alterations of activity, the caveat is relevant in the context of the distinction between direct and indirect drug effects on Behaviour.

A drug such as alcohol may very well modify neurotransmitter activity and some related Behaviour by actions on synthesis, degradation, receptor binding characteristics, or channel alteration. Nu merous other "side" effects might also occur. For example, ethanol consumption produces a marked alteration in general metabolism, producing among other things a general acidosis and increasing blood a-hydroxybutyrate levels. As â hydroxybutyrate is a usable energy source for brain tissue, marked changes in cerebral energy metabolism are likely to follow.

Such "indirect" effects on neural function, and perhaps Behaviour, would certainly appear subject to the same sort of heritable variation as other biochemical systems. Thus the degree of Behavioural modification by indirect as well as direct action of the pharmacological agent might also be influenced by genetic factors. While hoping not to be redundant, we believe the conclusion is obivous. That is, gene action may affect pharamcological responsiveness in a variety of ways other than via the specific neural system controlling the phenotype of interest. Keeping the foregoing caution in mind, numerous informative avenues of research on neurotransmitter system regulation by genetic variables are available.

The most well researched of these is probably the area of catecholamine neurotransmitters. Norepinephrine and dopamine systems are subject to functional alteration by numerous pharmacological agents; many of these agents are of use in a clinical setting precisely because of their Behavioural effects.

These functional alterations can occur as a result of inhibition or induction of the enzymes responsible for synthesis and degradation of these catecholamine transmitters. Early work by Ciaranello, Barchas, Kessler, and Barchas first established the fact of strain differences in brain activity of tyrosine hydroxylase (TH), the rate limiting enzyme in catecholamine biosynthesis.

The same report also demonstrated marked strain differences in adrenal phenylethanolamine-N methyl transferase (PNMT), the enzyme responsible for epinephrine production from norepinephrine. We here describe further examination of these mouse strain differences in TH activity in a context that serves as an example for pharmacogenetic study as well as neural development.

In midbrain of CBA/J mice, the activity of TH is 20per cent less than in BALB/cJ mice. A comparable difference in immunocyto chemically detectable dopamine neurons in midbrain suggests that this difference in TH activity is due to more cells with the enzyme, rather than to greater specific activity per cell. A correspondingly more dense innervation of the striatum, by what are presumably the axons of these midbrain nigral cells, occurs in the BALB/c mice.

It is likely that this difference is related in a general way to the fact that the BALB/c brain is simply a larger one. Nonetheless, this represents a clear example of genetic control over neuronal organization and neurochemical characteristics of a Behaviourally important system. The differences in density of striatal innervation apparently predispose a differential response to two agents with dopaminergic activity, amphetamine and apomorphine.

BALB/c mice are more responsive to the locomotor stimulant effects of amphetamine and less sensitive to the stereotypy producing effects of apomorphine, a dopamine receptor agonist. Interestingly, these charac teristics follow a developmental time course that lags slightly behind the appearance of the TH activity difference. Apparently both pre and postsynaptic functions of the dopamine system are different between the two strains as reflected in the actions of the two drugs.

A more detailed understanding of the genetic control over dopamine cell proliferation, growth, and effects on target tissue in this model system will answer many questions not only of pharmacogenetic interest, but also of basic importance to developmental neurobiology.

In this example of TH activity differences, the enzyme activity itself is probably not the focal point of either gene or pharmacological action. As noted, a more basic developmental issue is central. Yet the enzyme is a clear marker for a dramatic genetic influence. In other work, TH is seen to be more directly involved.

Gamma-butyrolactone (GBL), through its metabolite gamma-hydroxy butyric acid (GHB), produces an inhibition of impulse transmission in dopamine neurons. A compensatory activation of TH occurs in these cells and an accumulation of dopamine occurs. In the LS and SS mice, GLB has hypnotic effects that differ in the same direction as ethanol effects. Of interest here is the fact that dopamine accumulation is 50per cent greater in the LS mice after GBL treatment.

Presumably the activation of TH was greater in LS mice. This activation occurs via kinetic rather than molecular genetic induction and apparently reflects a differential effectiveness of GHB at a subcellular site in LS and SS dopamine neurons.

Although several studies of neurotransmitter systems' enzymes exist, there is yet to be a complete characterization of all biosynthetic and degradatory enzymes, transmitter levels and turnover rate, and receptor kinetics for any single system where genetic variation is also of interest. The difficulties of such a massive undertaking are obvious, but the benefits accruing to all three fields, neuropharmacology of drug action, neurochemical organization and dynamics, and mechanisms of genetic action are equally impressive. The field must surely progress along these lines.

ALCOHOL

The genetic effects on a given Behavioural response do not necessarily imply similar genetic effects on other phenotypes. The degree to which a similar genetic pattern occurs for more than one phenotype is a measure of the genetic correlation of these phenotypes. Yet taskor situation specific genetic effects are not rare. The LS and SS mice are usually very different in their response to ethanol.

Yet in one study, across a wide range of doses, a particular Behaviour was not differentially affected. When ethanol was used as a poison (UCS) in a conditioned taste aversion experiment, the magnitude of aversion produced was similar in the two lines. Thus, whatever systems underlie the loss of righting relfex differences do not control response to ethanol in the taste aversion experiment.

A second example may be drawn from the literature on strain differences in ethanol response. Although BALB/c mice lose the righting reflex longer than C_{57}BL/6 mice, they show a greater locomootor stimulation following low dose ethanol administration. It may be of interest that, in these two strains, amphetamine shows exactly opposite effects on locomotor activity. BALB/c mice are not activated, whereas C_{57}BL/10 mice are thus a negative genetic correlation is suggested for these effects of amphetamine and ethanol.

MEASURES OF RESPONSES TO OPIATES

It is becoming increasingly apparent that strain differences are readily obtained using the standard measures of opiate induced analgesia and locomotor activity. However, it is equally apparent that both the magnitude and the direction of these differences change with the particular measures and procedures employed. For example, $C_{57}BL$ mice have been characterized as being sensitive to morphine induced hyperactivity in open field tests and toggle boxes.

However, Eidelberg, Erspamer, Kreinick, and Harris have reported that $C_{57}BL$ mice respond to several doses of morphine with decreased activity when the measure is recorded on an activity wheel. Similarly, measures of analgesia, such as the hot plate technique, tail flick procedures, and abdominal writhing induced by agents such as acetic acid often yield very different strain distribution patterns. The exact nature of these differences among measures has been reviewed elsewhere.

For the present, the disparate results of these procedures lead us to argue for careful descriptions of the methods employed when general phenomena associated with opiate responsivity. However, we share the optimism of Crabbe and Belknap that consistent patterns among several dimensions of opiate responsiveness may yet emerge as the results of further investigations are known.

GENETIC CORRELATIONS

The basis of many inferences in pharmacogenetic research is the comparison of effects of more than one drug on the same genetic material. This is implicitly an example of genetic correlation. Both correlated drug effects and correlated effects of a single drug on several Behaviours are means for analyzing the domain of effects of that pharmacological agent, or the common physiological substrate of a variety of agents.

The quantitative genetic methods for deriving genetic correlations are readily available and are not presented in detail here. Instead, the major experimental approaches to genetic correlation methods and the general usefulness of such correlations.

One of the most utilized approaches to genetic correlations has been the examination of correlated characters in selected lines. A generalized example would find a set of lines selected for response to drug X, characterized at a variety of physiological levels, and perhaps other types of responses to the same drug being assessed as well. The usual argument is that care must be taken in interpretation of any correlated characters especially if replicated lines and unselected control lines are not available.

The basis of this assertion is that when only two lines are sampled and two characters are measured on those genotypes, then the degrees of freedom for assessing the correlation are $N - 2 = 0$. Two points are worthwhile in counter. A set of once replicated lines would produce a situation with $df = 2$, hardly a powerful statistical test.

Only when the numbers of genotypes exceeds 20 will actual correlations below.40 be detectable. In addition, correlational degrees of freedom are based on the assumption that both variables generating the covariance are free to vary. This is clearly not the case after a selection has been accomplished.

The selected phenotype is at that point fixed. Only the as yet unmeasured phenotype is "free" to vary. Perhaps a more Bayesian approach is desirable.

The most rational alternative to numerous replicates is a simultaneous measurement of several phenotypes during the course of selection. As selection progresses, those characters sharing a true genetic covariance with the selected phenotype will diverge in somewhat parallel fashion. As this approach forces one to identify relevant variables even before the selection begins, it is a less than perfect one.

Obviously a more practical approach is required. We would advocate a rational approach that makes use of a priori knowledge of the pharmacological system being examined and physiological intervention if the question being asked is one of mechanism of drug action. For example, in the LS and SS mice it is not surprising that alcohols other than ethanol have a differential effect.

However, if a correlated character is, for example, an altered preference for saccharin sweetened water, then one must wonder if drift is responsible. In sum, no single approach seems completely suitable. Therefore, only careful and objective scrutiny can avoid erroneous conclusions. An additional tool in genetic correlation analysis is the diallel cross and also the triple test cross. The inclusion of non homozygous animals in the diallel cross analysis is probably an advantage, but again, unless the number of genotypes is large, statistical power is low.

The use of genetic correlations to assess commonalities of drug action is a powerful approach. Only the use of cross tolerance studies can so clearly demon strate common substrates of action as can the presence of a genetic correlation. The aforementioned effects of higher order alcohols on LS and SS mice led to the conclusion that considerably different mechanisms of action underlie the effects of alcohols and other sedative hypnotics that did not differentially affect the two lines.

It is conceivable that, in a genetic system with high heritabilities, an experimenter could test a very large range of drug classes and attempt to classify them via one of a number of classification techniques on the basis of the genetic correlations.

The alternate approach examines the effects of a single drug on numerous Behaviours. Examination of the kinds of genotype patterns of Behaviours affected similarly by a given drug should indicate similar substrates for those Behaviours. It would also be of interest to examine the pattern of genetic correlations in drugged and undrugged mice separately. Changes in those correlations would suggest an ability of the drug to dissociate genetically based covariation of Behaviours. Thus even at a strictly Behavioural level of analysis, the genetic correlation could be a very useful tool.

DEVELOPMENTAL PHENOMENA

As already mentioned, the examination of drug responses in the developing organism and concomitant study of neural development is a powerful approach to the study of drug action and neural development. The addition of the genetic variable as a unique tool to tease apart neuropharmacological mechanisms is a potentially fruitful tactic. Correlated biochemical and pharmacological changes are more easily linked if the correlation holds in genotypes that show different time courses of developmental events. An example from the literature on alcohol preference in mice is instructive. Wood reviews this literature on age related differences in response to and preference for alcohol by laboratory rodents. Two studies in particular are instructive both for purposes of the genetic variable and its interaction with an experiential variable. BALB/c mice are, as adults, avoiders of ethanol.

Yet, from 4-9 weeks of age, they will voluntarily consume a fair amount of ethanol. Consumption diminished to near zero by 9-10 weeks unless a choice between alcohol and water was provided during the 4 to 8-week period. Interestingly, mice forced to drink alcohol during this period would typically not consume the alcohol as adults. Thus the fact of choice in young mice appears to predispose alcohol consumption in adults. This plasticity of phenotype or lack of phenostability makes clear the complex nature of the neural and Behavioural systems on which the genes can act only very indirectly.

A more general view of pharmacogenetic research must realise that a most critical variable in drug effectiveness is the functional organization of the complex target organ called the nervous system. It is ultimately the genetic influence on the assemblage of the brain and other physiological systems that provides the substrate for individual differences produced in drug responsivity.

A given drug acting on similar receptors in two brains that differ in organization of the neural network in which these receptors reside might produce different Behavioural outcomes. The mechanism of such gene effects on neural development may be as simple as cell proliferation and migration rates or as complex as regulation of the timing of critical and sensitive periods for Behavioural and emotional development.

The study of such biochemical characters as those already enumerated may yield important findings. Ultimately, however, a concern with developmental issues in pharmacogenetic work will yield much more information about the mechanisms of gene and drug action.

Chapter 6

Genetics of Schizophrenia

Schizophrenia is one of the more complex and less understood of the psychological disorders. For many years, studies have explored the genetic aspect of this syndrome in an effort to provide a better understanding of its etiology, symptoms, course, and development. Much of what is involved in schizophrenia still remains unclear, although great strides have been accomplished by research in the area.

In trying to understand the etiology of psychiatric disorders, one cannot avoid studying gene—environment interactions. Just as genes influence the susceptibility of an individual to the development of almost any disorder, environmental factors are also seen as important etiological contributors to disturbance. Consequently, whether or not genetic or environmental variation is involved in the development of schizophrenia is no longer considered to be an important point of contention in this area of research.

Instead, most of the current investigations appear to be concerned with acquiring a more thorough understanding of the magnitude and relevance of both environmental and genetic variation and the ways in which these interact to establish the various manifestations of schizophrenia.

Today there seems to be little doubt that a genetic basis for schizophrenia exists. However, the specific schizophrenia-producing genetic components and the method by which they are transmitted remain unclear. Genetic investigations have involved primarily three methodologies: family studies, twin studies, and adoption studies. Despite some difficulties, these methods have obtained strong evidence pointing toward the involvement of genetic factors in the development of schizophrenia. Currently, the amount of interest and speculation in this area of research is high, and on the basis of findings, sophisticated models and new, more innovative research techniques are being developed.

The major conceptual, diagnostic, and methodological difficulties inherent in this area of investigation, the evidence for the existence of genetic factors in the etiology of schizophrenia, and the models for a genetic mode of transmission that have arisen from these lines of research.

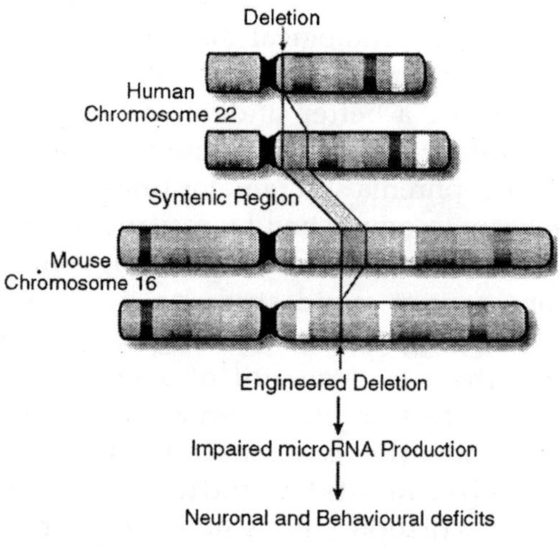

Fig. Genetics of Schizophrenia

One of the more difficult aspects of this field remains the lack of knowledge concerning the nature of schizophrenia. To date, no homogeneous group of schizophrenics has been isolated and no unitary etiology for the disorder has been found.

The possibility exists that schizophrenia is not a single, unitary disorder, but may consist of a number of different entities. Schizophrenic heterogeneity could be present at several levels. Seemingly similar phenotypic expressions may be due to quite different etiologies. As a result, the reliance upon symptomatology for the differentiation of schizophrenic subtypes may be grossly misleading. To complicate matters further, the complexity of gene operations makes it difficult to delineate the various interacting etiological components of a schizophrenic disorder.

The nature of the activating environment and the timing of that activation in addition to the gene could all combine to produce a single phenotype. Therefore, a range of possible phenotypic outcomes exists for every gene as a function of the activating environment and the time of activation. Depending on the environmental circumstances, there are a number of different genes that can produce a single phenotype. Consequently, several genetic mechanisms may underlie the expression of a single characteristic.

If schizophrenia is in fact heterogeneous, many of the present research strategies may prove inadequate because most rely on the implicit assumption of homogeneity among patients studied. A schizophrenic sample based on an overinclusive diagnostic category would result in variable schizophrenic performance. As a result, common findings among schizophrenics would be difficult to detect and valid hypotheses may be prematurely discarded on the basis that the data fail to meet statistical significance.

The failure to reach a clear consensus on the nature of schizophrenia has resulted in conflicting viewpoints. To date, genetic studies of schizophrenia have relied almost exclusively on the assumption of a model in which schizophrenia consists of a complex disorder amenable to the same type of study and displaying the same pathogenic characteristics as an inheritable disease.

Other points of view, especially those among the psychodynamic and contextualist schools of thought, hold that the conceptualization of schizophrenia as a disease entity is inadequate and unjustified. For example, Sarbin and Mancuso claim that schizophrenia is a label and the product of social processes resulting in the imposition of a moral verdict in response to certain undesirable Behaviours. Others have proposed that certain forms of familial interaction predispose an individual to schizophrenia, whereas still others have preferred to focus on the schizophrenic's intrapsychic world and dynamics. An interesting model postulated by Meehl incorporates both genetic and environmental factors to explain the occurrence of schizophrenia. Meehl suggests that all that is inherited in schizophrenia is an integrative neural defect that he labels *schizotaxia*.

The imposition of a social learning history on schizotaxic individuals results in a schizotypic personality organization. That is, all schizotaxic individuals obtain a schizotypic personality organization regardless of environmental conditions. Most of these individuals will remain compensated. However, those who encounter certain detrimental influences (e.g., schizophrenogenic mothering) will decompensate into schizophrenics. In light of this model, only a few of those individuals who have inherited the neural defect and are thereby predisposed to schizophrenia, will manifest the disorder.

Despite the wide disparity among theoretical models, most can be classified as elaborate extensions of the nature versus nurture controversy. It is apparent that this debate cannot be adequately resolved until a broader base of knowledge is developed.

Until that time, it may be wise to bear in mind that genetic research does not necessarily negate endeavors supporting contextualist and psychodynamic conceptions. Instead, genetic research addresses both environmental and genetic factors, and is interested in delineating the effects of each separately and in interaction as they pertain to schizophrenia.

DIAGNOSIS

Another research difficulty related to that of schizophrenic heterogeneity pertains to clinical diagnosis. The new DSM-III allows for considerable heterogeneity as to who will be diagnosed schizophrenic; persons need to fulfill only a certain number of criteria out of a standard set of possibilities in order to receive the diagnosis.

Schizophrenia cannot be diagnosed unless signs of the disorder have been present for at least 6 months. Illnesses of shorter duration, or illnesses that have resulted in the face of a psychosocial stressor and appear to be transient in nature are no longer included.

In fact, these disorders are nct even considered to be subtypes of schizophrenia. These changes in diagnostic criteria have left schizophrenia researchers in a peculiar bind since they now must try to define the onset of a schizophrenic disorder. In the past, the onset of schizophrenia was demarcated by the appearance of psychotic symptoms. However, it is well known that the onset of schizophrenia is often quite insidious, and that prodromal features commonly predate psychotic symptoms.

Logically, it would be best if we could define onset as the point in time when one could predict with virtual certainty that a schizophrenic psychosis would occur. Unfortunately, we still lack the necessary knowledge to make such a prediction, and the definition of onset remains strictly arbitrary. Consequently, the conceptual problem of trying to determine criteria for delineating signs along a continuum marking transitions from a schizophrenia-prone personality structure to a prodomal syndrome to a full-blown psychosis still exists.

Because of these diagnostic difficulties, a standard definition for what constitutes schizophrenia remains to be determined. The absence of a standard definition remains a major hindrance to the progression of this field. The severity of the problem is due to the fact that every research project must of necessity concern itself with identifying those people who have the disorder.

Therefore, the homogeneity of the sample studied and the rigor of the investigation carried out on the sample depend on the accuracy of the diagnosis. Unfortunately, there is no independent validation and only moderate reliability for diagnosing a schizophrenic disorder. Even the combination of moderate interrater reliability and independent validation appears to be inadequate. Consequently, there is no truly homogeneous sample in schizophrenia research. Instead, investigators are forced to contend with the heterogeneous groupings that fit the schizophrenic label, the abundance of contradictory data, and the numerous disagreements over research findings. As long as this state of affairs exists, studies are limited to statistical rather than clinical significance.

At present, many diagnostic difficulties result because little is known about the unity or diversity of schizophrenia. In order to gain a clearer understanding of the nature of the disorder, many investigations have been conducted to determine the degree to which the range of phenotypic expression in schizophrenia is paralleled by genetic heterogeneity. These included certain studies that were designed to investigate whether these subtypes were fundamentally alike or different.

One such study by Winokur, Morrison, Clancy, and Crowe found that hebephrenic probands had more than three times as many schizophrenic relatives as paranoid probands. These authors determined that both children and other secondary cases of schizophrenia in the families of hebephrenics were more likely to be diagnosed hebephrenic as opposed to paranoid schizophrenic. The same type of pattern was established for paranoid schizophrenics.

The children and other secondary cases of schizophrenia in the paranoid probands were more likely to manifest paranoid rather than hebephrenic symptoms. These familial data led the au thors to conclude that process schizophrenia is not likely to be a unitary illness. Instead, they postulated the existence of two types of schizophrenia in the population studied, one of which may manifest itself mostly as hebephrenia but also occasionally as paranoid schizophrenia.

The second type would manifest itself exclusively as paranoid schizophrenia in which there are relatively few schizophrenic relatives. Together, these findings suggest that there may be only one type of hebephrenic schizophrenia, but two types of paranoid schizophrenia, one related genetically to the hebephrenic subtype.

Another study by Kety et al. in 1975 demonstrated that the diagnoses of chronic, latent or borderline, and uncertain schizophrenia tend to cluster in the biological relatives of schizophrenic adoptees. Similarly, a study by Slater and Cowie examined the classical pre-Kraepelinian categories and found a significant association between schizophrenic probands and their schizophrenic offspring with respect to type of category.

Kringlen in 1968 corroborated these results and also demonstrated that the entire set of Kraepelinian subtypes is represented in the children of every type of schizophrenic proband.

These studies are typical of investigations in this area. Their results seem to indicate that some groups of disorders are genetically related, whereas the relationships among others remain unclear. Despite some contradictory findings, the most common results from these studies provide evidence for a common genetic basis among various forms of schizophrenic disorders.

In general, the findings indicate that cases of all subtypes occur in the families of each schizophrenic subtype. They also suggest that only a modest tendency toward subtype resemblance exists within families.

In addition to considering the unity or diversity of schizophrenia, one must also consider the relationship between schizophrenia and other psychiatric disturbances when making clinical diagnoses. This is especially important if one wants to formulate diagnostic criteria that are successful at differentiating among various psychoses.

At present little is known about the specificity of the schizophrenic disorder. However, there is some evidence suggesting a genetic relationship between schizophrenia and other psychiatric disorders. In a study by Reed, Hartley, Anderson, Phillips, and Johnson, a high risk of psychoses for the siblings of psychotics was found.

When the investigators looked at specific disorders, they discovered that the risk for first degree relatives of affective probands to develop a similar disorder was less than their risk of becoming schizophrenic. There is also some indication that the risk of schizophrenia for the offspring of paranoid and hebephrenic schizophrenics is lower than the risk for children of manic—depressives. Even some of the very earliest studies discovered that schizophrenic and affectively ill individuals were about equally prevalent in the families of schizophrenics. Today as well, despite efforts to select clinically homoge neous probands, the occurrence of different types of psychoses in the same family is difficult to avoid.

The suggestion that there may be a similar genetic basis to schizophrenia and other forms of psychoses raises the possibility of overlap in the symptomatology of various psychiatric disturbances; and indeed, this has been found to be the case. Another problem exists in the possibility that symptomatology is not directly related to genetic inheritance. Bleuler speculated that secondary symptoms such as delusions and hallucinations are not inherited. He also suggested that the primary symptoms do not correspond to the transmitted anomaly itself.

As a result of these possibilies, there are some major difficulties involved in basing diagnostic criteria on clinical symptoms. However, there are now some indications that future diagnostic categories will be based on biochemical indicators rather than Behavioural symptoms in order to decrease the heterogeneity of individuals in specified subgroups of psychopathology.

Using biological markers could be advantageous in several ways. First, it could avoid the genetically heterogeneous diagnostic groupings based on standard methods of symptom classification. It is well known that even relatively well formulated clinical subtypes of schizophrenia present diverse symptomatology.

As a result, it is impossible to differentiate between genotypic and phenotypic variations under the present diagnostic systems. The use of biologically independent variables in formulating diagnoses may promote the detection of biological etiologies and the formation of more homogeneous psychiatric diagnostic categories. Individuals who exhibit divergent or less severe clinical symptoms and are thereby excluded from present investigations could be included if they met a biological criterion.

This in turn would allow the high risk population (those

individuals with the salient factor) to be identified and studied. The use of a biologically based strategy would also permit the study of non hospitalized and unmedicated individuals, thereby avoiding the inclusion of these confounds within investigations.

In addition to developing biologically based criteria, many other efforts are being made to improve the diagnostic system. For example, the United States— United Kingdom Diagnostic Project, the International Pilot Study of Schizophrenia, and the World Health Organization have provided standardized input on patients for formulating diagnoses, experts to address diagnostic issues, and cross national reference points in order to establish criteria that will improve the reliability and validity of the diagnosis of schizophrenia on an international basis.

Additional reasons for optimism stem from the new trends in schizophrenia nosology. These include an increased reliance on affective symptoms as diagnostically useful, as well as a shift toward incorporating the schizo affective disorders within the affective diagnoses. In addition, the broad use of the schizophrenic label seems to be decreasing as a function of less emphasis being placed on classic schizophrenic symptoms, and more effort being devoted to the use of operationally defined criteria tested for reliability.

Although at present it is still difficult to reach clear and definite conclusions in research dealing with schizophrenia because of the many diagnostic difficulties associated with this disorder, the increasing refinement of diagnostic issues and the development of more sophisticated research techniques makes it possible to conceive of removing the current obstacles to gathering clinically significant data and conducting therapeutic—preventive research.

FAMILY STUDIES

Despite the numerous conceptual and diagnostic difficulties inherent in schizophrenia research, much evidence has been gathered to substantiate the existence of genetic components in the transmission of the disorder. The idea that various forms of mental illness are heritable has existed for many centuries. However, it was not until early in this century that researchers were able to provide actual documentation of this fact.

The first consanguinity study that found a significant increase in the prevalence rate of schizophrenia among relatives of index cases in comparison to the rate of the general population. More specifically, he discovered an age corrected risk for schizophrenia of 4.48per cent and 4.12per cent for other psychoses in the siblings of schizophrenics whose parents were unaffected.

Many other studies have confirmed Rudin's basic finding of a higher rate of schizophrenia in the relatives of samples of schizophrenics than in the samples from the general population. From these studies the rate of schizophrenia in the general population has been calculated to range from.35 to 2.85per cent , with a mean around.85per cent .

The findings from the consanguinity studies also reveal a clear correlation between the degree of blood relationship in relatives of an individual affected with schizophrenia and the risk of those relatives for becoming schizophrenic. More specifically, those relatives who share fewer genes in common with an affected individual show a lower expectancy rate for the disorder than those who share more genes.

In addition, these studies have demonstrated that there are no significant differences between prevalence rates for siblings and other first degree relatives. Although the difference found in the rate of schizophrenia between parents (5.5 per cent) and siblings and children of schizophrenic probands (10.2per cent and 13.9per cent respectively) appears to be large and therefore contradictory to the foregoing statement, this difference has been explained by the fact that these parents are largely selected for a degree of mental health through having been married and having children.

When a correction is made for this factor, the risk in parents of schizophrenics becomes close to that of siblings and children. The somewhat higher risk in children than in siblings of schizophrenics is thought to be due to sampling error.

Additional evidence for a genetic factor in the transmission of schizophrenia is found by comparing the different rates of prevalence for the disease among offspring of matings of various psychotic combinations of parents. Estimates have been obtained indicating that the risk for schizophrenia increases as the genetic loading increases. The best estimate of the risk of schizophrenia in the sibling of a schizophrenic is 9.7per cent when neither parent has the disorder.

This risk increases to 17.2per cent for the siblings of the proband when one parent also has schizophrenia. When both parents are schizophrenic, the risk to their children ranges between 36.6per cent and 46.3 per cent. Although it can be argued that this raise in risk from 17.2per cent to 46.3per cent for the siblings of a schizophrenic may be due to the extreme environmental conditions created by the mating of two schizophrenics, a study by Kallman reported that a comparable disturbed setting created by the mating of a schizophrenic and a psychopath yielded a sibling risk for schizophrenia of only 15per cent in Gottesman & Shields,. The basic findings of these studies have been further substantiated by the results of more recent investigations. For example, a study by Reed et al. reported that the risk to the offspring when neither parent was psychotic increased relative to the number of aunts and uncles who were psychiatrically disturbed.

The difference among estimates of risk is a fact significant for genetic theories. Not only does this difference demonstrate that the risk of mental illness increases as a function of increased genetic loading, but the large amount of variation in the offspring undermines the prediction of environmental theories that virtually all children from dual matings of schizophrenics would be emotionally disturbed.

Another line of evidence for the role of genetic factors in schizophrenia was provided by Manfred Bleuler, who applied the methodology of consanguinity studies to clinical symptomatology and course over time. He discovered that the biological relatives of schizophrenics who also had the disorder tended to have similar ages of onset, presenting clinical pictures, lengths of duration of illness, and outcomes. These findings can be interpreted in either an environmental or genetic fashion. However, a genetic explanation appears to carry more weight because many of the relatives in the study were geographically separated and in some cases did not even know each other. Therefore, the findings of the study indicate that genetic factors play a role in several clinical parameters of schizophrenia.

Although these family studies have contributed much toward the understanding of schizophrenia and have pointed to the role of genetic factors in the development of this disorder, some major difficulties are inherent in their methodology. The earliest family studies did not question the assumption that schizophrenia is heritable and therefore failed to consider the possibility that environmental factors such as family interactions also contribute to the etiology of the disorder.

As a result, these studies provided sources of data on the empirical risk of the occurrence of schizophrenia in different kinds of relatives, but did not elucidate the specific etiology involved in creating those risks. In other words, these studies could indicate if a trait ran in families, but could not specify what aspect of the family determined the trait. Obviously, families share more than genetic similarity, they also share a variety of social, psychological, and cultural experiences. Consequently, the results of these studies can just as validly be interpreted as supporting an environmental position.

Recently, some studies taking an environmental approach have investigated family interactions on the premise that certain conditions of family life can predispose an individual to schizophrenia. The majority of these investigations have focused on deviant role relationships and disordered communication processes among family members.

The findings of these studies generally indicate that deficits in family communication are highly correlated with schizophrenia and high risks for schizophrenia in offspring. In addition, there are some indications that families of schizophrenics contain poorly articulated and rigid role structures in the areas of dominance, autonomy, and sexuality. Despite the success of this type of research, it is unable to establish the occurrence of schizophrenia as a strictly environmental outcome.

In fact, the evidence that is presently available tends to suggest that both genetics and family process play a role in the development of schizophrenia. In light of this evidence, the current mode of investigation has changed to incorporate biological, genetic, psychological, and social factors in the context of complex, multifactorial models of causation when approaching family variables in relation to schizophrenia.

TWIN STUDIES

It was not until the advent of twin studies that the role of genetic and environmental factors in schizophrenia could be more clearly defined. The power of the twin method resides in the environmental similarities and genetic differences between monozygotic and dizygotic twins. Both types of twins share the same intrauterine environment simultaneously with their co- twin.

However, monozygotic twins share the same fertilized ovum initially, whereas dizygotic twins are no more genetically alike than ordinary siblings. As a result, one can assume the presence of a genetic factor in determining a trait if the concordance rate for that particular trait is significantly higher for monozygotic as opposed to dizygotic twins.

The methodology of twin studies also provides certain advantages. For example, through the comparison of intrapair resemblance, it becomes possible to identify subtypes or components of schizophrenia that may be more under genetic or environmental control than others. Also, by noting the variability of abnormal personality traits in the monozygotic co-twins of typical schizophrenics, it becomes feasible to identify schizophrenic "equivalents" or schizotypes.

The comparison of monozygotic twins who differ in outcome will permit the formulation of inference about the relationship between life experiences and outcome. It must be kept in mind that the twin study method relies on certain assumptions. These specify that monozygotic twins as such are not specifically predisposed to schizophrenia, and that the environments of monozygotic twins are not systematically more similar than those of dizygotic twins in such a way as to foster the disorder.

Also, when assessing the outcome of these studies it is important to be aware of the usage of statistical age corrections and their effects on research findings. Age corrections incorporate estimates of a lifetime risk for developing schizophrenia in order to compensate for a current sample while the disorder exhibits a variable age of onset. Although several methods have been developed, none is entirely satisfactory, and all present only varying degrees of approximation to the true concordance rate.

It is essential to realise that the absolute size of the concordance rate is not predictive of the relative importance of the genetic contribution to the trait in question. Concordance rates can be expressed in several fashions. The basic difference in methods depends on whether concordance is expressed in terms of the pair or the individual.

The casewise or proband method refers to the proportion of cases with an affected partner, whereas the pairwise rate considers the proportion of pairs in which both twins are affected. The individual or proband concordance method yields a higher percentage than a pairwise concordance. Therefore, only those studies that use the same method are comparable.

The most influential early twin studies were conducted by Kallman and Slater. Even though the two studies contained important differences and suffered from methodological problems, both found a significantly higher concordance rate for schizophrenia among monozygotic twins.

By the end of Kallman's study and with some diagnostic revision, but without statistical age correction, 59per cent of 174 monozygotic pairs and 9per cent of 517 dizygotic pairs were concordant for what Kallman labeled "definite schizophrenia." In Slater's study, 65per cent of 37 monozygotic pairs and 14per cent of 58 same sexed dizygotic pairs were found to be concordant. Both studies used the pairwise method for calculating the concordance rates. Findings from other earlier studies were basically consistent with these results.

More recently, several studies have been conducted that are methodologically powerful, and that provide highly significant results. One of these studies, reported by Tienari, was conducted on an exclusively male sample in Finland. Initially, this investigator found a zero concordance rate for monozygotic twins.

However, by 1976 the pairwise concordance rate for the sample had risen to 16 per cent. The use of the proband method, the concordance rate for the sample was found to be between 35per cent and 53per cent depending upon which borderline cases were included. As a result, even though it took some time, a significant monozygotic—dizygotic difference was demonstrated.

A study by Kringlen was conducted in Norway and included both male and female twins. His study was especially important for several reasons. His sample was well constructed and contained the largest number of psychotic twins personally investigated among the more recent studies. His investigation was very comprehensive and included case history, pedigree, clinical analysis, rating scale data, and all functional psychoses in addition to schizophrenia. Using the more conservative pairwise method, Kringlen found a significant difference between monozygotic and dizygotic twins with concordance rates ranging from 60per cent for monozygotic probands rated as severe to 25per cent for the most benign.

His results indicated no difference between the concordance rates of monozygotic twins as a function of gender. He reported that whether the sexes of the dizygotic twins were the same or different, the concordance rates did not differ, but were equal tc those found in ordinary siblings. These results counter findings from some earlier studies and undermine the psychodynamic thesis that concordance for schizophrenia will increase without regard to genetic similarity, but relative to the amount of likeness and resulting identity confusion among siblings. The findings of another important twin study were reported by Fischer, Harvald, and Hauge in 1969. This study was conducted in Denmark upon a sample that was very representative with respect to twin types and sexes of probands. Only those probands who met strict criteria for chronic or process schizophrenia were included.

Even so, schizophreniform, paranoid, and atypical psychoses in co-twins led to data analysis for quantitative concordance for other than process schizophrenia. As a result of this study, Fischer et al. discovered a pairwise concordance rate of 24per cent for monozygotic twins and 10per cent for dizygotic twins.

In addition, the authors calculated the rates according to the proband method and obtained a 36per cent concordance rate for monozygotic and 18per cent concordance rate for dizygotic pairs. When those co-twins who exhibited a form of schizophrenic disorder different from process schizophrenia were included in their calculations, the investigators obtained proband concordance rates of 56per cent and 26per cent for monozygotic and dizygotic twins respectively.

These figures appear to be good approximations of the true incidence rate, as the twins used in this study were followed long enough to virtually eliminate the need for any statistical age correction.

Fisher went on to report that monozygotic twin pairs concordant for schizophrenia were not at an increased risk for the disorder when compared to discordant monozygotic twins. She did not discover a relationship between severity of disturbance in the proband and the risk of developing psychosis in the co-twin.

The offspring of concordant twins were not at an increased risk for schizophrenia when compared to the offspring of discordant twins. In addition, the likelihood of a discordant twin member producing a schizophrenic child was equal to that of the schizophrenic member of the twin pair. These findings point toward a genetic hypothesis and indicate that the genetic transmission of schizophrenia is not always manifested directly.

The Maudsley Hospital based study of Gottesman and Shields conducted in England obtained similar results to those of the studies reported earlier. These investigators also obtained a significantly increased monozygotic concordance rate over that found in dizygotic twins. Similarly, Pollin, Allen, Hoffer, Stabenau, and Hrubec, in a study based on an American sample, found a higher concordance rate for monozygotic as opposed to dizygotic twins. More specifically, the concordance rate for schizophrenia in monozygotic twins was found to be 3.3 times greater than the rate in dizygotic twins.

What is interesting about this study is that the results were based on an exclusively male sample consisting of almost 16,000 twin pairs who had successfully passed army induction exams and who had served in the armed forces during the period spanning World War II and the Korean War.

The calculations for the concordance rates were based on chart diagnoses obtained from VA records or responses to mailed questionnaires. As a result, the consistency of the findings from this study appear somewhat amazing when one considered the size of the sample, its apparent skewness toward healthy individuals, and the tendency for wartime nonpsychiatric physicians to use the diagnosis of schizophrenia more loosely.

This last factor tends to lower the homogeneity of the schizophrenic sample and the accompanying concordance rates for schizophrenia. Nevertheless, a monozygotic to dizygotic twin concordance ratio of 3.3 was obtained, and the authors concluded that the genetic loading for schizophrenia was higher than for any other psychiatric or medical diagnosis considered in their investigation.

At present, all the major twin studies carried out worldwide have demonstrated higher concordance rates for monozygotic as opposed to dizygotic twins. The monozygotic concordance rates from studies conducted before 1965 tended to average around 60 per cent or more, whereas the results from more recent studies are usually lower. These differences have been attributed to variations in methodological and diagnostic procedures. Even so, the consistency of twin studies' findings despite diversity in design and execution provides strong evidence for a hereditary factor in the etiology of schizophrenia.

From these studies it is apparent that identical twins per se are not more prone to develop the disease. However, if one twin does develop the disorder, the co-twin is at an increased risk over an ordinary sibling or dizygotic twin. In addition, a frequent finding from twin studies has been that concordance is higher for index cases with severe forms of schizophrenia.

The findings from these studies contradict the suggestion that the stress inherent in rearing twins plays a role in the development of schizophrenia by demonstrating that both monozygotic and dizygotic twin pairs are not at an increased risk for the disorder. In fact, the research literature in general suggests that life events and stress do not account for the development of a schizophrenic disorder.

Despite the usefulness of the twin method in demonstrating that genetic factors play a role in the etiology of schizophrenia, it is still important to consider the ways in which environmental factors may have contributed to the findings reported. Jackson suggested that monozygotic twins may be more prone to develop schizophrenia because of identity confusion and/or weak identity formation.

However, this idea has failed to be substantiated, as no excess of monozygotic twins has been discovered in schizophrenic samples. The possibility of one twin identifying with the other, in this case disturbed twin, has also been considered as a psychological factor that might lead to higher concordance rates in monozygotic twins. Usually this is referred to as *folie à deux*, or psychosis by association, a clinical entity that would result in clinical resemblances between the psychoses of identical twins.

Again, there appears to be a lack of evidence to support this position. There are simply too few cases of shared delusions or other symptoms of intense identification of one schizophrenic twin with the other to support the theory.

Another environmental explanation that has been put

forward postulates that both members of monozygotic twin pairs have a greater likelihood of being exposed to a schizophrenia inducing situation than do members of dizygotic twin pairs. This position would entail the assumption that monozygotic twins are treated more alike than dizygotic twins.

In fact, Kallman reported higher concordance rates in twin pairs living together for 5 or more years before the onset of the disorder than in pairs separated during that period. Even so, studies conducted on identical twins where one or both members later became schizophrenic reveal similar concordance rates for those reared apart when compared to those reared together. In addition, pairs of monozygotic twins reared apart reveal much higher concordance rates for schizophrenia than dizygotic twins reared together.

One criticism of the twin method addresses the assumption that the sharing of the intrauterine environment simultaneously by both members of the monozygotic twin pair markedly reduces the developmental and environmental variations for the members of the pair. Instead, it is often the case that one twin member is put at a disadvantage and receives less of the limited space and nutrients available than the other twin. Consequently, the disadvantaged twin may appear weaker and receive special care and attention from the parents. This parental Behaviour may in turn foster dependency and submissiveness in the individual.

In a study conducted by Pollin et al. 1966, the conclusion was reached that low birth weight was an important contributing factor to the development of the schizophrenia. However, this finding did not appear to be replicated in other studies. For example, Kringlen found no general tendency for submissive, mother dependent twins to become schizophrenic. In fact, his data suggest that birth weight and obstetric complications are not correlated with the later development of schizophrenia.

Instead, he concluded that the twin who is predisposed to become schizophrenic, is predisposed in a nonspecific manner. The conclusions in this area remain unclear however.

Other more recent analyses reveal that the schizophrenic twin is often lighter at birth in discordant twin pairs. Even so, the differences in birth weights are usually small, and the schizophrenic twin's weight is usually within the normal range.

Despite the continuing controversy in this matter, it is important to remember that even if some conclusive evidence could be obtained with regard to birth weights, it would still throw little light on the etiology of schizophrenic disorders because lowered birth weights can result from numerous factors and may create any number of reactions that may or may not be etiologically linked to schizophrenia.

Taken together, the findings from the twin study method have strengthened the view that genetic factors are significantly involved in creating the higher monozygotic concordance rates reported in the literature. This is not to say that environmental variables are not important in the etiology of schizophrenia. However, the findings do point out the vital contribution twin studies have made for the genetic position in this area of research.

ADOPTION STUDIES

In the adoption strategy there are basically four approaches that can be used to separate genetic factors from other possible etiological factors related to familial Behaviour and environment the adoptees' families' method, schizophrenic adoptees are selected as index cases to be compared to a control group. The control group consists of other adoptees who are free of any psychiatric history but who match the index cases with respect to age, sex, age at transfer to adoptive parents, and social class of adoptive parents. The difference in incidence of schizophrenia is then determined for both the biological and adoptive relatives of the index and control groups.

The adoptees' study method consists of finding schizophrenics whose children were placed for adoption at an early age so as to compare the prevalence ratings of these children with either a matched control group or the population in general. A third strategy, the adoptive parents' method, starts with the adoptee schizophrenic index cases and then compares the incidence of pathology in their biological and adoptive parents.

Cross fostering is perhaps the most sophisticated research strategy among adoption studies. With this research design two different groups of adoptees are identified: those who have a biological parent who is schizophrenic, but whose adoptive parents are free of psychiatric illness, and those whose biological parents have no psychiatric history, but who are adopted by a schizophrenic. The two groups are then compared with respect to incidence of schizophrenia and other psychiatric disturbances.

One study that uses the adoptees' method was conducted by Heston. He found that the offspring of severely schizophrenic mothers who were given up for adoption within the first 3 days of life still grew up to have schizophrenia at the same rate as expected for those reared by their schizophrenic mothers. In addition, the offspring from schizophrenic mothers showed considerably more abnormalities other than schizophrenia than would be expected for the general population. A later and more elaborate study conducted on a Danish population confirmed Heston's results.

Another Danish study conducted by Kety, Rosenthal, Wender, & Schulsinger identified adoptees who became schizophrenic and then through the use of documents determined the status of their biological and adoptive parents, siblings, and half siblings. The purpose of the study was to discover if the schizophrenia in the adoptees was a function of schizophrenia in the adoptive or biological families. Again, supporting evidence was found for a genetic factor in the transmission of schizophrenia.

It is important to note that Rosenthal used a broad diagnostic concept, the "schizophrenic spectrum" in their investigations. Rosenthal included borderline and schizoid or paranoid diagnoses along with the diagnosis of schizophrenia; and Kety included chronic, acute, or borderline schizophrenia, definite or uncertain, in addition to inadequate personality in the schizophrenic spectrum. Both studies reported that schizophrenic spectrum index cases showed an increase in the rate of schizophrenic spectrum disorders only in their biological relatives. Besides indicating that schizophrenia is heritable, these studies also offer evidence that the genetic potential for schizophrenia may manifest itself in ways other than overt schizophrenia.

Since these findings were reported, Kety has conducted interviews with the relatives of schizophrenics and the control adoptees in the field, in addition to increasing his sample size. Similarly, Rosenthal increased his sample size of index adoptees and interviewed the spouses of the schizophrenic parents. In general, the more recent data from these studies confirmed the findings of the earlier reports. More specifically, there were more schizophrenic and schizophrenia related disorders in the biological relatives of schizophrenic adoptees and in the adopted away children of schizophrenic parents than in the control groups.

The cross fostering technique, also found a greater prevalence of psychopathology among the adopted away offspring of schizophrenics, but no such increase among the cross fostered group. The investigators concluded that genetic factors play a role in the etiology of schizophrenia whereas familial psychopathology does not. An earlier study conducted by Wender, Rosenthal, and Kety looked at the parents of adult schizophrenics who had been given up for adoption as infants, and compared them with parents who had reared their own, now adult schizophrenic offspring.

These two groups were also compared to the adopting parents of normal young adults. The findings of this research indicated that the biological parents of schizophrenics were considerably more disturbed than the adopting parents of schizophrenics, and that the adopting parents of schizophrenics were exhibiting a slightly greater degree of psychopathology than the adopting parents of normal young adults.

Consequently, both the psychosocial and genetic

hypotheses were supported by this research. In a later study based on the original test data, protocols were blindly analyzed, and both the adoptive and biological parents of schizophrenic patients were successfully differentiated from the adoptive parents of normal offspring. An earlier series of experiments using projective techniques in blind predictions of psychiatric patients and their families was also completed.

In addition to matching the tests of the patients with their families, test data from other members of the family were successfully employed to deduce the diagnosis, form of thinking, and degree of disorganization in the young adult psychiatric patient. These findings provided substantial support for the environmentalist position and seemed to indicate that certain types of familial communication patterns are possible etiological contributors to schizophrenia. The debate between those supporting a contextualist position and those espousing a genetic model raged on, as so often happens in this area of investigation, without arriving at a firm resolution.

Part of the difficulty in interpreting this type of research stems from the inability to avoid genetic and psychological confounds and the failure to differentiate schizophrenogenic Behaviour from the Behaviour of schizophrenics. In addition, problems arise when the findings from certain studies may be attributable to several hypotheses. For example, the Wynne et al. results may just as easily be attributed to psychological changes in the parents resulting from rearing a schizophrenic child, or genetic factors accountable for both the psychopathology and communication patterns observed.

That is, even if communication deviances are etiologically significant for schizophrenia, they do not rule out the possible important of genetic factors as influential in creating those deviances. In any case, further supportive evidence for a genetic position was obtained in a later replicative study by Wender, Rosenthal, Rainer, Greenhill, and Sarlin using the same design but including a systematic sample. The findings from this study revealed that the biological parents of schizophrenics were more disturbed than adoptive parents of either schizophrenics or normals; and that the adoptive parents did not differ on degree of psychopathology.

Despite all the advances made by the adoption research, it is still important to realise that even the most sophisticated adoption studies are limited in their resolving power. As of yet, confounds and other research difficulties cannot be avoided. This is in part due to the many practical considerations that must be dealt with when using this type of research strategy.

For instance, only certain children are permitted to be legally adopted, and sometimes these are abnormal before placement. Adopting parents as a group usually create a restricted range of rearing environments and share a similar socioeconomic status with the biological parents. In addition, fostered children are not included in the initial sample; and the spouses of schizophrenics whose children are adopted probably include a higher proportion of socially inadequate individuals than those spouses of schizophrenics whose children are not adopted.

Although these limitations do not erode the principal conclusions obtained through adoption studies, they do limit their generalizability. In spite of the various difficulties inherent in this research method, the adoption research has made a vital contribution to the investigation of the etiology of schizophrenia. Perhaps most importantly, it has disconfirmed the hypothesis that the high rate of schizophrenia observed in the offspring of schizophrenics is necessarily due to being raised by a schizophrenic parent.

As a result of their findings, the adoption studies along with the family and twin studies have all but dispelled any doubt that a genetic factor is operating in the transmission of schizophrenia. This seems pretty remarkable when one considers that this genetic factor has surfaced in spite of all the problems in the formulation of individual diagnoses, the heterogeneous nature of the schizophrenic syndrome, and the interindividual differences that must be present in any population labeled schizophrenic.

With the knowledge that hereditary factors play an

important role in the development of schizophrenia, scientists have begun to search for what exactly is transmitted in the disorder. However, so far the search for an adequate model has not yielded conclusive results. What has happened, is that investigators have proposed several theoretical models and have explored the fit between the theoretical predictions made from these models and the existing data.

The proposed models for the genetic transmission of schizophrenia can be roughly divided into two types: monogenic and polygenic. Monogenic models rest on the assumption that the action of a single or a few major genes leads to a specific metabolic deficit that in turn causes schizophrenia. Polygenic models propose that the action of many genes, each with a small effect, can, given a certain combination, produce the disorder.

Since Rudin conducted his study in 1916, it has been apparent that the mode of genetic transmission in schizophrenia does not seem to follow any simple Mendelian pattern. In all genetic studies, the number of cases of schizophrenia found in the relatives of schizophrenic probands has always been less, but never higher than would be expected with a simple Mendelian distribution.

Therefore, one can conclude that the incidence of schizophrenia was not due to random error nor could it be explained by a simple monogenic theory. Consequently, investigators have had to propose the existence of modifying factors to accompany monogenic theories in order to account for the lack of consistency with Mendelian ratios.

In 1965, Slater proposed a major partially dominant gene model that included the concept of penetrance. Penetrance refers to the percentage of cases that manifest the trait in question when the necessary genetic components are present. According to Slater's modified dominance theory, virtually all schizophrenics inherit the same abnormal gene in a single or double dose, but only one out of four will manifest the disorder, because the gene expression is only partially penetrant. Later, modifications of this model by other investigators such as Slater and Cowie and Kidd and Cavalli

Sforza produced variations in the posited penetrance of the dominant gene in the heterozygote.

Another model developed by Kallman in 1953 proposes a recessive single gene that can be modified by another variable called the "constitutional resistance to manifestation" This variable was considered to be a polygenetically determined trait. Those individuals who were homozygous with respect to the single gene were thought to be much more likely to develop schizophrenia than those individuals who were heterozygous. However, the degree of constitutional resistance present could affect the tendency to develop schizophrenia even when the individual was homozygous.

Through this model, which allows for various combinations of homozygosity and heterozygosity interacting with high or low levels of constitutional resistance, Kallman hoped to account for different degrees of severity in the manifestation of the disorder.

One criticism that has been leveled against these monogenic models is that they cannot stand without some kind of support from such concepts as penetrance or resistance to manifestation. These types of concepts are especially problematic because the specific mechanisms proposed have never been demonstrated, do not further the understanding of the etiology of schizophrenia, and fail to provide a means of disproving a particular model. Consequently, these types of concepts have often been labeled as ad hoc explanations whose only purpose is to bring order and consistency to a theoretical model.

When one works with polygenic models many of the difficulties inherent in the monogenic models disappear but others arise. Mendelian distributions are no longer required and the expectancy rates for schizophrenia, which were lower than expected for monogenic positions, are not contraindicative of polygenic theory. There is good scientific precedent for this type of theory, as many traits among many different kinds of species have been shown to have complex non Mendelian, multifactorial genetic determinants.

One interesting model advanced by Falconer uses a

quantitative genetic approach and makes the assumption that underlying the etiology of the disorder is a variable distributed along a continuum. This variable, called liability, includes both innate tendencies and external circumstances that may be predisposing to the development of the disorder. The variation of liability is distributed in the form of a normal, bell shaped curve.

At one end of this curve is a point called the threshold, which demarcates those individuals who are affected by the disorder from those who are not. This point is positioned on the curve according to the prevalence of the disorder in the general population. Once the prevalence rates among different classes of relatives of affected individuals are determined, it becomes possible to use the model to calculate the heritability of the liability of the disorder in question.

In 1967, Gottesman and Shields were the first to apply Falconer's model to schizophrenia. They obtained many heritability values using different samples of subjects, and these values ranged between 45 and 106 per cent. Accordingly, the authors acknowledged that the method was subject to errors, and that the underlying assumptions of the model had not all been fully met.

Even so, the authors noted that the findings did indicate the heritability of liability to schizophrenia across samples to be substantial. However, they also qualified these findings by pointing out that heritability is not a property of the trait alone, but also of the population sampled and of environmental circumstances.

Another polygenic theory has been proposed by Cancro. His model assumes the existence of a continuum with regard to phenotypic expression in the schizophrenic syndrome, and does not assume that the phenotype is inherently abnormal. That is, different subgroups of schizophrenia are assumed to be involved in different combinations or patterns of phenotypes. In addition, there are a number of different genes that can produce each phenotype depending on the environmental conditions present. Therefore, two individuals who share the same genetically loaded trait may have obtained that characteristic by different means. Obviously, Cancro's model is quite complex. Not only does he have the possibility of a particular phenotype developing from different pathways, he also allows for the heterogeneity of the schizophrenic disorder by assuming that multiple traits are involved. Although these polygenic models contain several advantages over the monogenic models, they are also more loosely formulated and consequently more difficult to test.

Studies measuring the fit of theoretical models to the ever increasing amount of data are continually taking place. For instance, Kidd and Cavalli Sforza reviewed the then current research data and tested the goodness of fit of these data to both a monogenic and polygenic model, using a threshold approach. On the basis of their analysis, neither a monogenic nor polygenic model could be discarded, as both seemed to fit the available data equally well. This led the authors to speculate that any intermediate model between the two extremes investigated would also fit the data satisfactorily. In fact, they postulated that any model of genetic heterogeneity would fit, with improvement in fit increasing as more parameters are included in the calculations.

Interestingly, when Mathysse and Kidd expanded on this approach, they found that neither of the two extreme genetic hypotheses were adequate when trying to account for the observed data. The single major locus or monogenic model provided estimates that were too low, whereas the multifactorial model provided estimates that were too high relative to the observed incidences of schizophrenia among monozygotic twins, and offspring of parents who were both affected.

Only a model with two interacting genes could possibly account for the incidences observed in the input data. Even so, both the monogenic and multifactorial models did predict genetic heterogeneity among schizophrenics and provided some interesting perspectives on the disorder.

When the input data were fitted according to the monogenic model, the genetic composition varied along with the frequency of the schizophrenia producing allele. Every individual has two alleles or genes, one from each parent, at each locus. At low frequencies of the allele, the development of the disorder seems to be largely due to environmental factors. At relatively high frequencies of the allele, the proportion of heterozygotes increases whereas the number of phenocopies decreases.

The proportion of homozygotes remains low no matter what the allelic frequency. Consequently, the single major locus model predicts that genes play an extremely important role on those occasions when they are present in double dose, but those cases would be rare even among persons manifesting the disorder.

Under the polygenic model, however, Mathysse and Kidd found that 9.1 per cent of the schizophrenic population has such a high genetic risk that environmental factors play virtually no role in the development of the disorder. That is, this model proposes that in one of every 11 schizophrenics, the disorder is almost totally genetically determined. Obviously the results are still unclear and the conclusions remain tentative. Arguments over genetic hypotheses will probably continue to lack meaning as long as virtually all models can account for the data. Until the research data becomes homogeneous, future investigations will be unable to rely on the use of statistical techniques alone to clarify which genetic model is most applicable to schizophrenia.

Although monogenic and polygenic models represent the genetic extremes of inheritance, there are also models of genetic heterogeneity that rest on the principle that more than one gene at one locus is needed to produce schizophrenia. Although appearing similar to polygenic theory, the two theories are different because polygenic models require that more than one locus is necessary for the development of the disorder.

The effects of different schizophrenia producing genes are not cumulative in genetic heterogeneity theory as they are in polygenic models, and the heterogeneity model does not propose a continuum. Instead, each gene is believed to be separate and distinct in its effects. Consequently, a particular combination of genes is not required to produce the disorder, but rather the appropriate specific genes. A heterogeneity model had been proposed by Karlsson. He postulated a two gene theory in which a mechanism of modified dominance was involved. The theory allowed for the prediction of distinctive personality types from specific genotypes. However, there does not appear to have been any independent validation of this model.

Taking a different approach to trying to uncover the mode of inheritance in schizophrenia, Mitsuda examined different types of schizophrenics and then classified them according to outcome. Cases for the apparent mode of transmission were classified into three groups: dominant, recessive, and intermediate. His findings suggested that the dominant gene group contained more prognostically favorable cases, whereas the recessive group had more that were unfavorable, and the intermediate group fell somewhere in between. As a result, he decided on the existence of a heterogeneity model for which three modes of genetic transmission were possible.

At present, theories of heterogeneity are very popular. However, if they are correct, they imply that the likelihood of discovering the mode of genetic transmission in schizophrenia is very low. On the other hand, monogenic theories appear more inviting because they allow for the possibility of identifying exactly what is transmitted in schizophrenia, thereby offering hope that therapeutic and preventive techniques will be developed sometime in the future. In light of these differences, it may be more worthwhile to direct research toward trying to disprove the single mode theory, if only to provide some direction as to which model is "true."

It is apparent that no single model for the mode of genetic transmission in schizophrenia has been universally accepted. There are even disagreements among investigators espousing the same type of model. For example, among the monogenic theorists there is no clear consensus regarding the mode of action of a single gene. However, there does seem to be agreement among the research results that a relatively low frequency of homozygotes exists in the population, and that the manifestation rate is very high for these individuals. At present, it is still not possible to decide on exactly what is inherited in schizophrenia and how it is transmitted.

However, the studies conducted to date, along with the improvements for research strategies that they foster, provide the necessary impetus for more sophisticated model building and testing. These in turn, provide reason for optimism that this area of investigation is worthwhile, and that it will lead to the eventual discovery of the mode of transmission in schizophrenia.

METHODOLOGICAL DIFFICULTIES

There are many methods that have been employed in the study of genetics and schizophrenia. Twin studies have been especially valuable for assessing the importance of biological factors in an illness. Of course, their value is still dependent on the validity of the diagnoses of zygosity and disorder. Despite their usefulness, they still present several problems for data collection and analysis. In particular, the usage of different definitions of concordance rates can be particularly problematic.

Monozygotic twins often differ on biological traits. Other factors such as obstetrical complications and different birth weights may interact with susceptibility to psychiatric disorders such as schizophrenia.

Separation studies, including studies of adopted children, cross fostering studies, studies of half siblings, and studies of monozygotic twins reared apart, have proven invaluable by demonstrating unequivocably that genetic factors are involved in the transmission of schizophrenia. Even so, these studies in themselves cannot elucidate what genetic and biological factors are operating and how they contribute to the development of a schizophrenic disorder.

Instead, the best contemporary use of separation studies may be to shed more light on the relative importance and specificity of certain environmental factors in relation to schizophrenia. This could possibly be accomplished through the comparison of rearing environments with populations at high and low genetic risk. In this manner, a negative correlation between genetic risk and a specific environmental factor would suggest a causal role for that factor.

Nuclear families have constituted the most popular form of data for genetic studies. This type of data is easily related to the epidemiology of the disease, but often poses difficulties for genetic analyses and interpretations. Part of the difficulty stems from the researchers' failure to distinguish the parent—offspring and sibling—sibling relationships. Instead, risks are calculated only for first degree relatives, while the finer distinctions are ignored.

Failure to note the sex of both proband and relative is also common and may affect morbid risk calculations. Until these factors are taken into account and sufficient specifications by relationship are made, these types of studies will remain limited in utility and ability to discriminate between proposed genetic models and types of heterogeneity.

Another valuable but often neglected resource in the investigation of genetic factors in schizophrenia is the use of multigenerational pedigrees. One major advantage of this approach is that it allows for the potential identification of homogeneous phenotypic subgroups on the basis that any rare disorder found in several members of one family is likely to have the same etiology in all. Pedigrees are amenable to the study of genetic linkage and the testing of genetic models.

Sampling procedures also pose major problems in this area of research. For example, Lewine, Watt, and Grubb point out that the current use of parent—child concordance rates may result in the use of a sample that consists of schizophrenics with an unusually strong genetic and environmental predisposition to the disorder. These authors have also demonstrated how the overrepresen tation of females among index parents may introduce systematic sex differences in schizophrenic characteristics as well as biased prevalence rates.

This in turn may have negative implications for the interpretations based on this type of research. For example, current risk studies rely on theoretical concepts and assessment procedures based for the most part on adult male populations.

At present, it is still unknown how generalizable these conceptualizations are to female schizophrenics. It is difficult to ascertain if findings regarding the offspring of schizophrenic women are comparable to the findings for offspring of schizophrenic men.

Although we know that the risk for schizophrenia is the same for both men and women, the evidence indicating that schizophrenia may take different forms in men and women suggests that different etiological factors may be of importance for the sexes. The preponderance of a reactive, atypical, or schizo affective type of schizophrenia in women suggests that careful diagnoses of index parents and offspring according to schizophrenic subtype are necessary to clarify the nature of schizophrenia, including its possible heterogeneity. In addition, the fact that varying sampling procedures may result in differing concordance rates does not permit easy comparison of results among studies.

The importance of studying families over unrelated individuals seems to be well established. As such, it appears to be the optimal strategy for genetic research on schizophrenia at this time. In addition, it is apparent that many unknowns remain in this area of investigation. Therefore, until such a time as our methods of analyses and diagnostic capabilities have improved sufficiently, one needs to remain aware of the tentativeness of our present conclusions as to the exact nature and transmission of a genetic factor in schizophrenia.

Anytime one delves into the schizophrenia literature, one is struck by the overwhelming amount of information available in this area. Given the wealth of knowledge that has accumulated in this field, it seems almost ironic that schizophrenia remains one of the most complex and least understood of all the psychological disorders.

It is obvious that none of the theoretical models proposed to date is sufficiently capable of explaining schizophrenia. Each one of them is lacking in certain respects. For example, most research studies dealing with genetic or family theories cannot explain the occurrence of schizophrenia as either a strictly environmental or genetic outcome. Instead, the evidence tends to suggest that both genetics and family process play a role in the development of the disorder. Similarly, although biochemical and neurophysiological theories hold much promise with regard to deriving useful diagnostic categories, their focus appears too narrow to explain the full scope of the schizophrenic syndrome.

On the other hand, psychological and psychoanalytic theories appear comprehensive in their provision of meaningful ways for conceptualizing and attending to schizophrenic process. Because no one can perceive the schizophrenic's subjective experience, these theories appear to try to understand its nature and etiology through an intuitive or empathic approach. Unfortunately, these theories are difficult to investigate and therefore it is difficult to obtain enough empirical data to support these positions.

In trying to compensate for these deficits, much of the emphasis in research designs has been placed on analysis and description leading to a wealth of confusing and often contradictory data. This condition appears to be typical of the early stages of development of any complex field of knowledge; accordingly, many would argue that a need for synthesis that would integrate apparently opposing theories now exists.

However, a synthesis of incomplete information would not necessarily lead to a better understanding of schizophrenia. By ignoring some of the dichotomies and polarities in the data, by narrowing the focus to trying to find the interconnections among the information presently available, some valuable information might be lost. At the same time, one wants to avoid becoming engulfed in spending too much time and effort in the analysis of small details, as is often the case with research designed to support a particular theory.

It all boils down to the dilemma of not being able to formulate a proper integrative principle or theoretical model until one has at least enough information to delineate the concept one is involved with, and one cannot obtain that information unless a great deal of varied research encompassing many different areas of investigation is conducted. In order to reach the goal of a more comprehensive theoretical framework for schizophrenia, it seems important to do several things.

First, it appears vital that theory, research, and intervention remain interconnected so that any discovery in one area will influence the other. At present, dangerous rifts among these areas appear to be developing, wherein practitioners apply interventions with little or no theoretical basis and theoreticians form postulates without any specific delineation for integration into treatment or research strategies.

Similarly, we cannot afford to dismiss any of the material presently available, but instead should try to appraise and evaluate it objectively without any preconceived ideas or bias stemming from our own theoretical models.

Chapter 7

Mendel's Laws of Genetic

The discoveries that Mendel made with peas have been found to apply everywhere throughout the plant and animal kingdoms—to flowering plants, to mosses, to insects, snails, crustacea, fishes, amphibians, birds, and mammals (including man).

There must be something that these widely separated groups of plants and animals have in common—some simple mechanism perhaps—to give such definite and orderly series or results. There is, in fact, a mechanism, possessed alike by animals and plants, that fulfils the requirements of Mendel's principles.

THE CELLULAR BASIS OF HEREDITY AND DEVELOPMENT

In order to appreciate the full force of the evidence, a few familiar facts, that became known before the discovery of the mechanism in question, may be briefly reviewed. Throughout the greater part of the last century, while students of evolution and of heredity were engaged in what may be called the more general aspects of the subject, there existed another group of students who were engaged in working out the minute structure of the material basis of the living organism.

They found that organs such as the brain, the heart, the liver, the lungs, the kidneys, etc., are not themselves the units of structure, but that all these organs can be reduced to a simpler unit that repeats itself a thousand fold in every organ. We call this unit a cell.

The egg is a cell, and the spermatozoon is a cell.

Fertilization is the union of two cells. Simple as the process of fertilization appears to us today, its discovery swept aside a vast amount of mystical speculation concerning the rôle of the male and of the female in the act of procreation. Within the cell a new microcosm was revealed. Every cell was found to contain a spherical body called the nucleus. Within the nucleus is a network of fibres; a sap fills the interstices of the network.

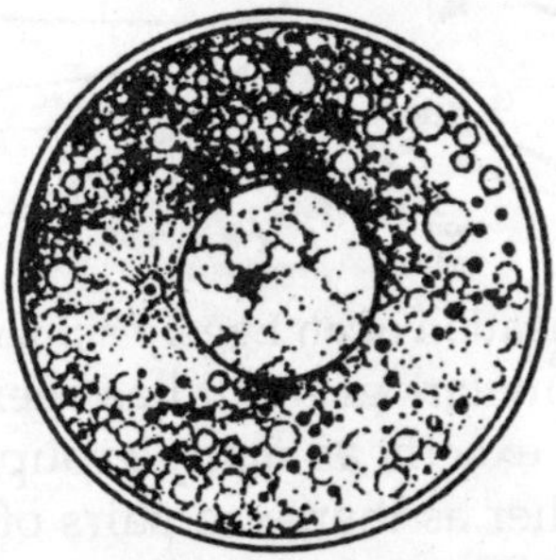

Fig. Cytoplasm and Centrosome with Astral Rays

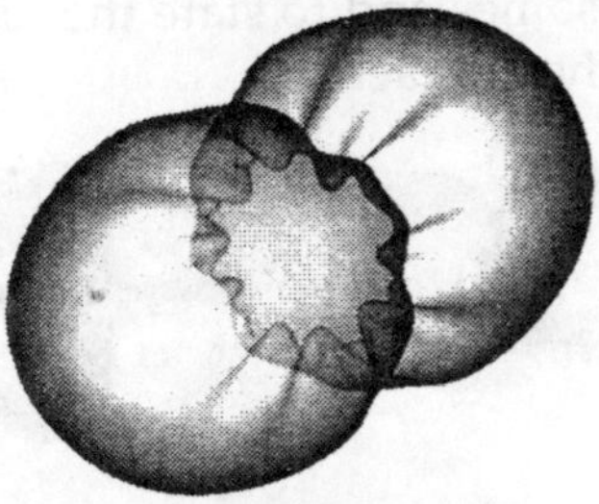

Fig. Cell Division

The network resolves itself into a definite number of threads or rods at each division of the cell. These rods we call chromosomes. Each species of animals and plants possesses a characteristic number of chromosomes which have a definite size, and sometimes a specific shape, and even characteristic granules at different levels. Beyond this point our strongest microscopes fail to penetrate.

Observation has reached, for the time being, its limit. Certain evidence relating to inheritance through the sperm led to the conclusion that the chromosomes are the bearers of the hereditary units. If so, there should be many such units carried by *each* chromosome; for, the number of chromosomes is limited while the number of independently inherited characters is large.

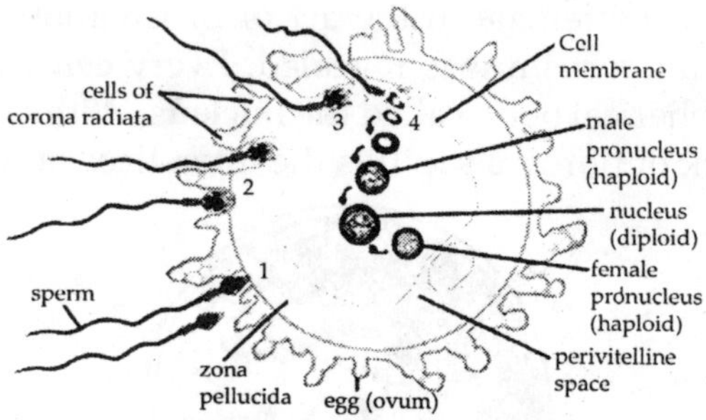

Fig. Fertilization of an Egg by a Spermatozoon

In Drosophila melanogaster it has been demonstrated not only that there are exactly as many groups of characters that are inherited together as there are pairs of chromosomes, but even that it is possible to locate the hereditary elements in particular chromosomes and to state the *relative position* there of the factors for the characters.

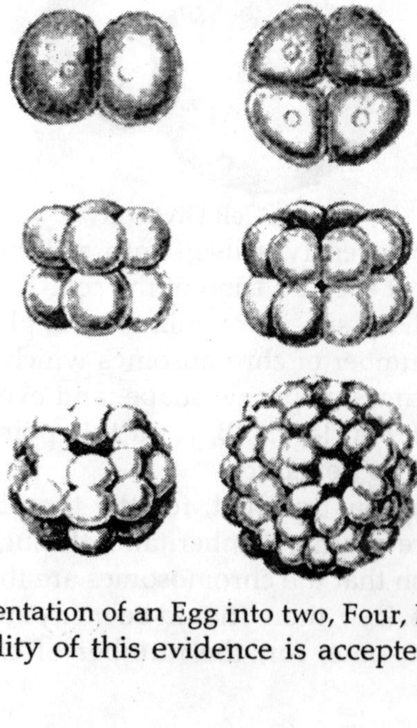

Fig. The Segmentation of an Egg into two, Four, Eight Cells, etc

If the validity of this evidence is accepted, the study of

the cell leads to the ultimate units about which the whole process of the transmission of the hereditary factors turns. Before considering this somewhat technical matter, certain facts, which are familiar for the most part, should be recalled, because, on these, rests the whole of the subsequent explanation.

The thousands of cells that make up the cell state that we call an animal or plant come from the fertilized egg. An hour or two after fertilization the egg divides into two cells. Then each half divides again.

Each quarter next divides. The process continues until a large number of cells is formed and, out of these, organs mold themselves. At every division of the cell the chromosomes also divide. Half of these have come from the mother, half from the father. Every cell contains, therefore, the sum total of all the chromosomes, and if these are the bearers of the hereditary qualities, every cell in the body, whatever its function, has a common inheritance.

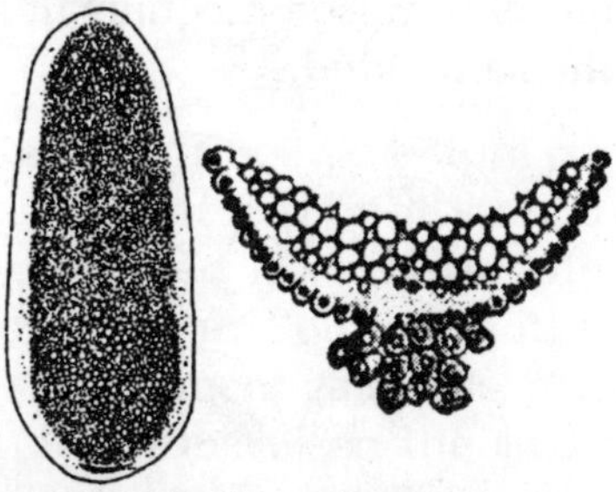

Fig. a, Section of egg of Calligrapha bigsbyana, showing "germ cell determinants" (granules), *g c d*, at posterior end of egg; *b*, posterior end of a later stage of same, showing primordial germ cells; *c*, Section of egg of Miastor, showing single primordial germ cell at posterior end

At an early stage in the development of the animal certain cells are set apart to form the organs of reproduction. In some animals these cells can be identified early in the cleavage.

The reproductive cells are at first like all the other cells in the body in that they contain a full complement of chromosomes, half paternal and half maternal in origin. They divide as do the other cells of the body for a long time. At each division each chromosome splits lengthwise and its halves migrate to opposite poles of the spindle.

But there comes a time when a new process appears in the germ cells. It is essentially the same in the egg and in the spermcells.

The discovery of this process we owe to the laborious researches of many workers in many countries. The chromosomes come together in pairs.

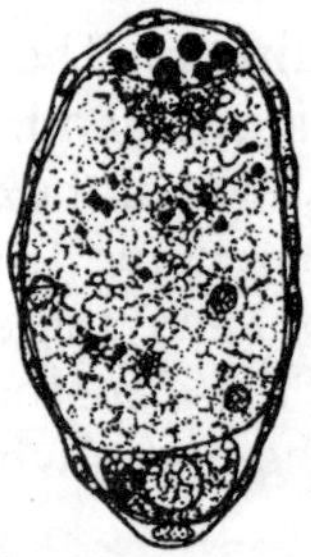

Each maternal chromosome conjugates with a paternal chromosome of the same kind.

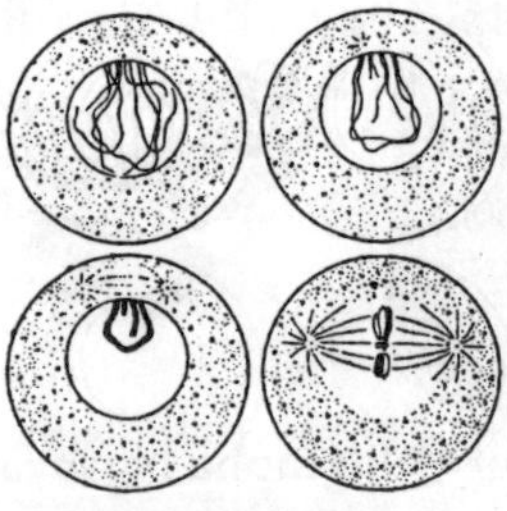

Fig. Two Maturation Divisions of the Germ Cells in the Male

These conjugate in pairs, *b*, beginning at the two ends of each loop. The threads contract, and a spindle appears, *d*, near the nucleus. The conjugating chromosomes enter the spindle, *d*. There they separate, *e*, moving to opposite poles of the spindle. The cell protoplasm begins to constrict, *f*. The chromosomes may without entering upon a resting nuclear stage pass onto a new spindle that has developed by the division of each of the centrosomes of each daughter cell, *g*. Each chromosome now splits throughout its length (equational division); half of each goes to one or the other pole. The two daughter cells then divide, giving four cells, each of which differentiates into a spermatozoon. Then follow two rapid divisions.

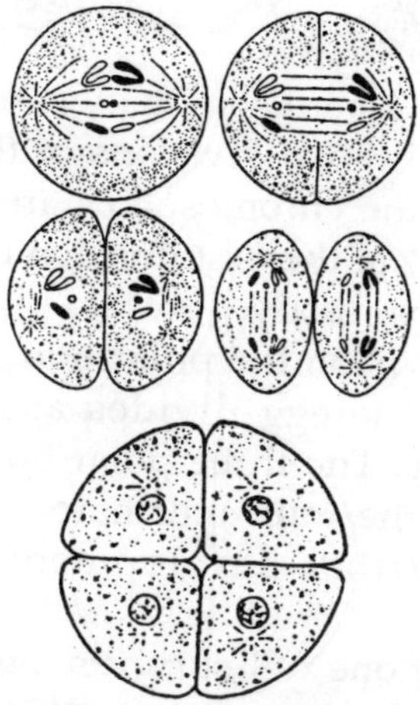

At one of the divisions the double chromosomes separate so that each resulting cell comes to contain some maternal and some paternal chromosomes, i.e., one or the other member of each pair. At the other division each chromosome simply splits as in ordinary cell division. In the male four spermatozoa are produced (by these two divisions) from each cell of the testis.

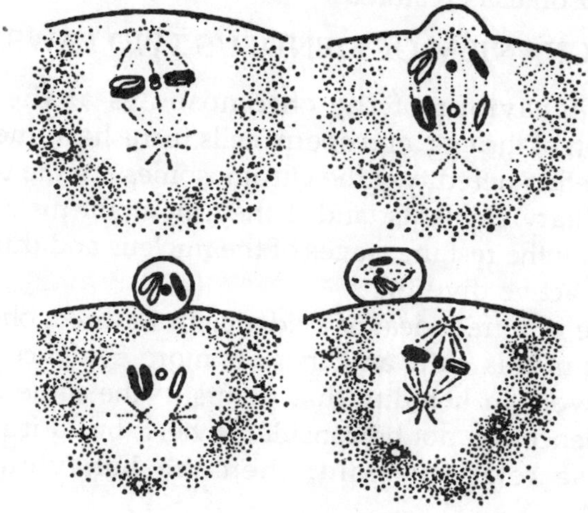

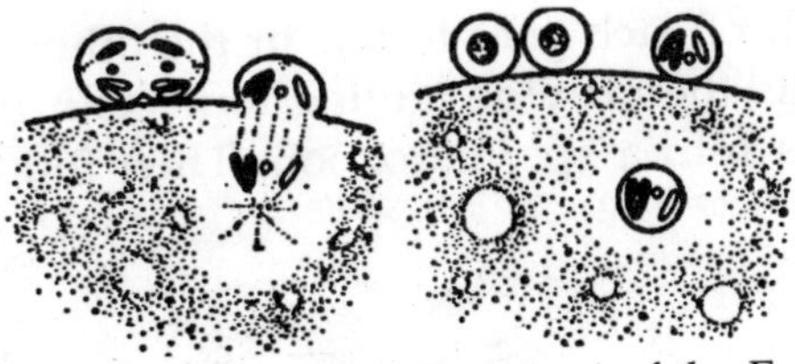

Fig. Two Maturation Divisions of the Egg

In the female the two divisions of the egg cell are very unequal, although the chromosomes are distributed equally to. all the cells. Thus at the first division one cell is very small and is called the polar body.

At the next division the polar body divides again, and at the same time the egg divides again also, producing another polar body. The three polar bodies and the eggcell are equivalent to the four spermatozoa, but only the egg cell undergoes further development—the polar bodies disappear.

Although only one cell survives, nevertheless there will be as many kinds of mature eggs as there are kinds of sperm cells (with respect to the distribution of the chromosomes), if, as we now know to be the case, the distribution of the chromosomes in the two final divisions (maturation divisions) are the same in the eggs and in the sperm cells. When the eggs are fertilized, each by one spermatozoon, the whole number of chromosomes is restored.

THE MECHANISM OF MENDEL'S TWO LAWS

The behaviour of the chromosomes at the time of maturation of the egg and sperm cells furnishes a mechanism for Mendelian heredity if the chromosomes are the bearers of the hereditary elements, and if they maintain their integrity both during the rest ing stages of the nucleus and during their period of active division.

There is a great deal of evidence from direct observation in favour of this view and there is more evidence from the modern work in heredity that points in the same direction. This evidence can not be considered here, but if it is granted that these relations hold, then the behaviour of the chromosomes during maturation furnishes, as stated above, an explanation of Mendel's laws.

If in the four o'clock the elements for red flower colour are carried in the red parent by the two members of the same pair of chromosomes and the elements for white flower colour are carried in the white parent by two members of the same pair of chromosomes, the germ cells (ripe egg and sperm cells) will each carry one of these chromosomes. If the red plant is crossed to the white, the pink hybrid will have a red and a white bearing chromosome.

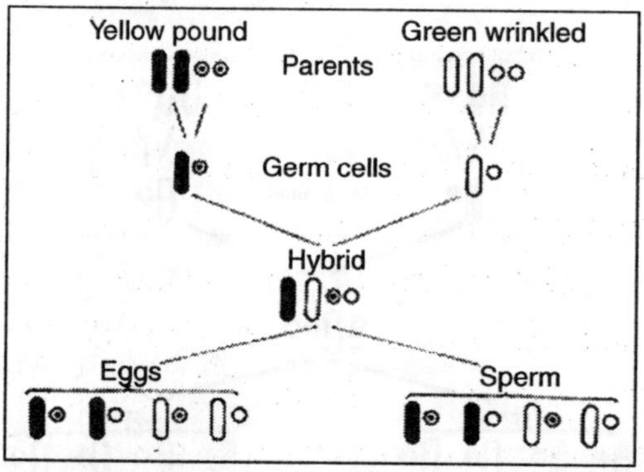

Fig. Diagram to Illustrate the Distribution of the Chromosomes in a Cross Between a red and a White Four o'Clock

When in the hybrid the germ cells ripen, these two chromosomes, being mates, will come together as a pair and then separate at one of the two maturation divisions, and half of the eggs will contain the red bearing chromosome and half will contain the white bearing chromosome. Similarly for the pollen grains. Chance fertilization of any egg by any sperm will give the combinations of chromosomes that Mendel's law of segregation requires. In other words the known behaviour of the chromosomes is exactly the same as Mendel's postulated elements.

Mendel's second law for the inheritance for two or more characters also finds its explanation in the behaviour of the chromosomes, provided the members of the pairs of chromosomes are sorted out independently of each other. For example, in the cross between yellow round and green wrinkled peas, if one pair of chromosomes in the hybrid (F_1) carries the contrasted elements yellow and green and another pair of chromosomes of the same hybrid carries the round and wrinkled elements, then, if these chromosomes at the maturation period behave independently, there will be four kinds of germ cells produced.

These four kinds will carry a yellow bearing and a round bearing chromosome, or a yellow bearing and a wrinkled bearing chromosome, or a greenbearing and a round bearing chromosome, or a green bearing and a wrinkled bearing chromosome.

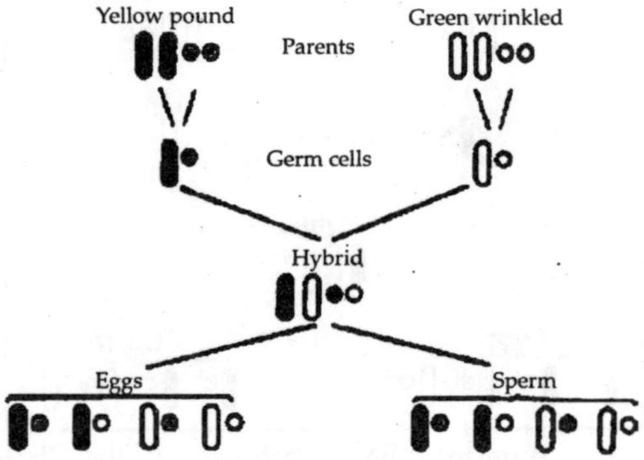

Fig. Two Pairs of Chromosomes Carrying two Pairs of Mendelian Factors

Only these four kinds of germ cells are possible on the chromosome mechanism. Self fertilization of such a hybrid will give the same recombinations of chromosomes that Mendel's second law requires for the hereditary elements.

THE LINKAGE GROUPS AND THE CHROMOSOMES

IF the hereditary elements, the genes, are carried by the chromosomes and if the chromosomes are persisting structures, there should be as many groups of hereditary characters as there are kinds of chromosomes. In only a few cases has a sufficient number of characters been studied to show whether there is any correspondence between the number of hereditary groups of characters and the number of chromosomes.

In the vinegar fly, Drosophila, there are about four hundred characters that fall into four groups. The characters are arranged in four groups, Group I, II, III and IV. Three of these groups are equally large or nearly so; Group IV contains only three characters. The characters are put into these groups because, in heredity, the members of each group tend to be inherited together, i.e., if two or more enter the cross together they tend to remain together through subsequent generations. On the other hand, any member of one group is inherited entirely independently of any member of the other groups; in the same way as Mendel's yellow green pair of characters is inherited independently of the round wrinkled pair.

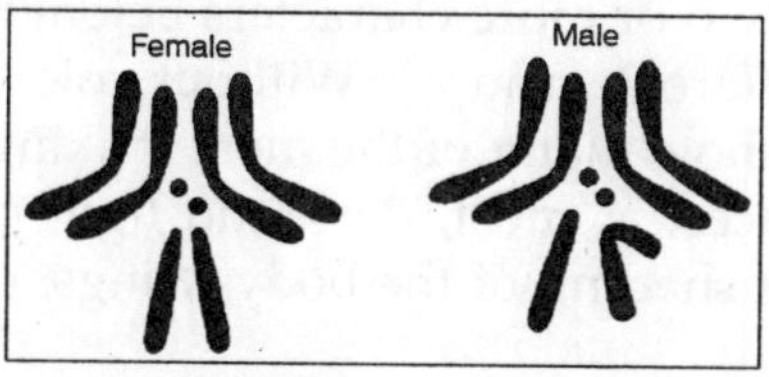

Fig. Female Group and Male Group of Chromosomes of Drosophila Melanogaster

In the chromosome group of Drosophila melanogaster there are *four* pairs of chromosomes, three of nearly the same size and one much smaller. Not only is there agreement between the number of hereditary group and the number of the chromo somes, but even the size relations are the same, for there are three large groups of characters and three pairs of large chromosomes, and one small group of characters and one pair of small chromosomes.

THE FOUR LINKAGE GROUPS OF DROSOPHILA MELANOGASTER

The following description of the characters of the wild fly may be useful in connection with the account of the modifications of these characters that appear in the mutants. The head and thorax of the wild fly are grayishyellow, the abdomen is banded wit alternate stripes of yellow and black. In the male, there are three narrow bands and a black tip. In the female there are five black bands. The wings are gray with a surface texture of such a kind that at certain angles they are iridescent. The eyes are a deep, brick red. The minute hairs that cover the body have a characteristic arrangement that is most obvious on the head and thorax. There is a definite number of larger hairs called bristles which have a characteristic position and are used for diagnostic purposes in classifying the species. On the foreleg of the male there is a comb like organ formed by a row of bristles; it is absent in the female. The comb is a secondary sexual character.

This is done to economize space, but of course there would be no difficulty in actually bringing together in the same individual any two or more characters belonging to the same group (or to different groups). Without colored figures it is not possible to show many of the most striking differences of these mutant races; at most, dark and light coloring can be indicated by the shading of the body, wings, or eyes.

GROUP I

The hereditary elements of this group are carried by the X-chromosomes. The characters are said to be sex linked. In the six flies drawn there are shown five different wing characters. The first of these types (*a*) is called cut, because the ends of the wings look as though they had been cut to a point. The antennae are displaced downward and appressed and their bristle like aristae are crumpled.

The second figure (*b*) represents a fly with a notch in the ends of the wings. This character is dominant, but the same factor that produces the notch in the wings is also a recessive lethal factor; because of this latter effect of the character, no males of this race exist, and the females of the race are never pure but hybrid. This same figure (*b*) is used here to show two other sex linked characters. The spines on the thorax are twisted or kinky, which is due to a factor called "forked." The effect is best seen on the thorax, but all spines on the body are similarly modified; even the minute hairs are also affected. The lighter colour of the body and antennae is intended to indicate that the character tan is also present. The tan flies are interesting because they have lost the positive heliotropism. As this peculiarity of the tan flies is inherited like all the other sexlinked characters, it follows that when a tan female is bred to a wild male all the sons inherit the recessive tan colour and indifference to light, while the daughters show the dominant sex linked character of their father, i.e., they are "gray," and go to the light. Hence when such a brood is disturbed the females fly to the light, but the males remain behind.

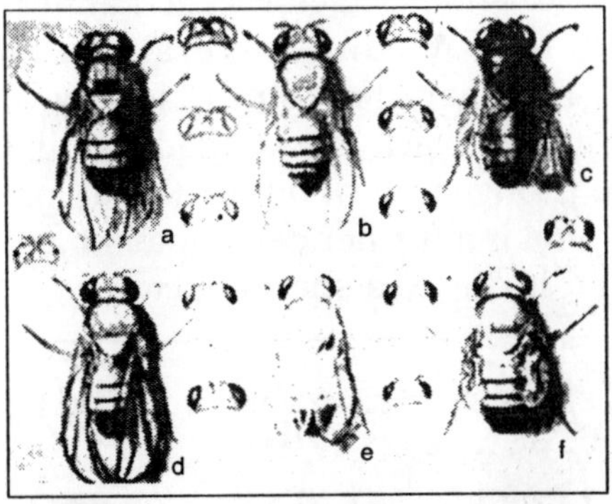

Fig. Some of the Characters of the first Chromosome of Drosophila Melanogaster

One of the first mutants that appeared was called rudimentary on account of the condition of the wings (*c*). Not all the flies of this stock have the wings in this condition; some have fully expanded wings that appear normal in all respects.

Nevertheless, about the same percentage of offspring show the pads irrespective of whether the parents had pads or expanded wings. The flies of this stock show, however, another character, which is a product of the same factor, and which is constant, i.e., repeated in all individuals. The two bristles on the sides of the thorax are constantly absent in this race. There are many different eye colors in Group I, ranging from a pure white eye to vermilion, which could be shown only by the use of colored drawings.

GROUP II

The hereditary elements of members of Group II are carried by one pair of the two large bent chromosomes.

In the first drawing (*a*) of which contains members of Group II, the wings are almost entirely absent or "vestigial." This condition arose at a single step and breeds true, although it appears to be influenced to some extent by temperature, also by modifiers that sometimes appear in the stock. The mutant is called jaunty.

The wing is long and narrow and sometimes bent back on itself. In several respects the wing resembles strap but seems to be due to another factor, called antlered. The wings are long and narrow and several of the veins are unrepresented. This character, "strap," is very variable. On the thorax there is a deep black mark called trefoil. In the wild fly there is a three pronged mark on the thorax present in many individuals. Trefoil is a further development and modification of this mark and is due to a special factor.

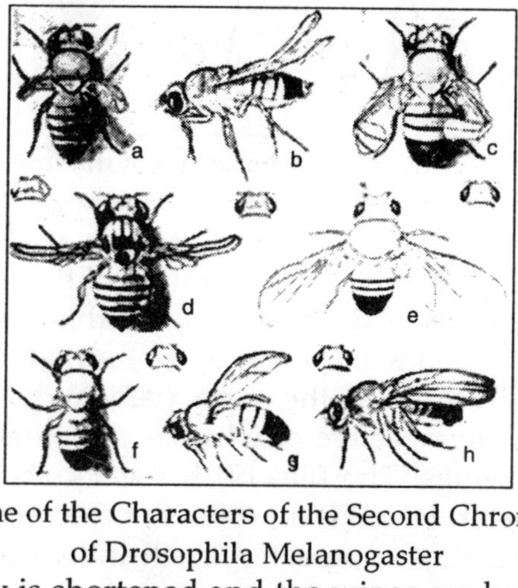

Fig. Some of the Characters of the Second Chromosome of Drosophila Melanogaster

The body is shortened and the wings are broad and held out. The legs are short. The absence of cross veins in the wings is characteristic. In the sixth figure (*f*), apterous, the wings are entirely absent, not even the base remaining as in vestigial. The apterous flies are almost completely sterile. The seventh figure (*g*) shows the wings "curved." In addition there is present a minute black speck at the base of each wing, due to another factor called speck. In the eighth figure (*h*) the wings are arched. The factor is called arc. The dark colour of the body, and especially of the wings, indicates the factor for black.

There are also a number of different eye colors in this group—one of which, brown, is darker in old flies than the red of the wild type.

GROUP III

The hereditary elements of Group III are carried by the other pair of large bent chromosomes. The old metathorax is replaced by another mesothorax thrust in between the normal mesothorax and the abdomen. It carries a pair of wings that do not completely unfold. On this new mesothorax the characteristic arrangement of the bristles is shown. Thus at a single step a typical region of the body has doubled. The character is recessive.

In the second figure (*b*) the dark colour of the fly is due to a factor called ebony. The size of adult flies varies according to the amount of nourishment obtained by the larva. After the fly emerges its size remains nearly constant, as in many insects. Two races have, however, been separated that are different in size as a result of a genetic factor. The first of these, called dwarf, is represented (*c*). The race is small but variable in size, depending on food and other conditions. The same figure shows the presence of another factor, "sooty," that makes the fly dark.

In the fourth figure (*d*) another mutation in size is shown. It is called "giant." The flies are twice the size of wild flies. In the fifth figure (*e*) the mutant dichaete is shown. It is characterized by the absence of two of the bristles on the thorax. Other bristles may also be absent, but not so constantly as the two just mentioned.

Another effect of the same factor is the spread out condition of the wings. In the sixth figure (*f*) the wings are curled up over the back; the character is called curled. In the seventh figure (*g*) the wings are beaded, i.e., the margin is defective at intervals, giving a beaded like outline to the wings. This condition is very variable and much affected by other factors that influence the shape of the wings. There are many eye colors in this Group—one of these, sepia, becomes very dark in old flies.

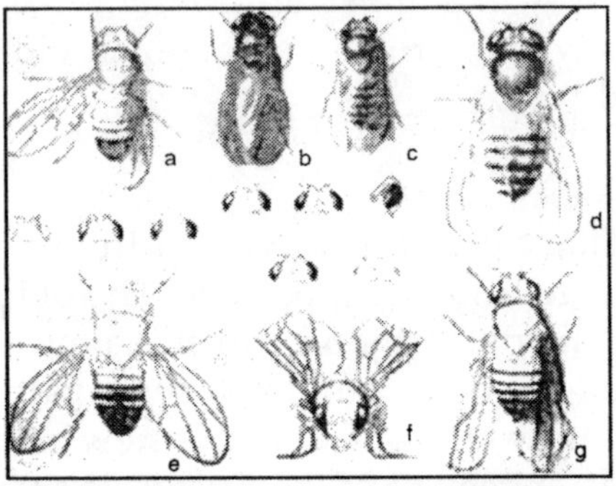

Fig. Some of the Characters of the third Chromosome of Drosophila Melanogaster

Pink and peach eye colors are modifications of the same gene (a case of multiple allelomorphs). Two other eye colors in this group, scarlet and cardinal, are almost indistinguishable, but the genes for these characters lie in quite different parts of the chromosome.

GROUP IV

The hereditary elements of Group IV are carried by the pair of very small chromosomes. Only three mutants have been obtained. One of these, called "eyeless", is variable—the eyes are often entirely absent or represented by a few or by several ommatidia (b, b^1). On the sides of the head where the normal eye lies there is, in "eyeless," a corresponding empty area in the more extreme condition (a, a^1), and even when a piece of the eye is present it lies in this area but failing to fill all of it, the outline of the full sized eye is, so to speak, still present. These parts of eyes (b, b^1) might be spoken of as rudimentary organs.

Drawing (*c*), represents "bent," so called from the shape of the wings. This mutant is likewise very variable, often indistinguishable from the wild type, yet when well developed strikingly different from any other mutant. The third mutant (*d*) is called shaven. The bristles and hairs are extremely short, and the thorax especially, appears as though shaven.

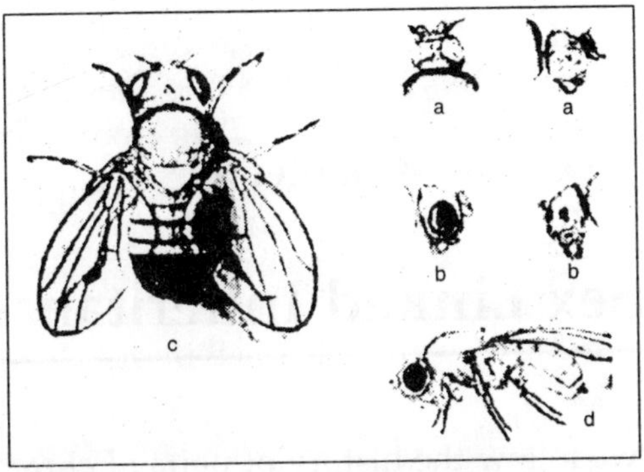

Fig. The Three Characters of the Fourth Chromosome of Drosophila Melanogaster

This brief account of a few of the mutant races that can be most easily represented by uncolored figures will serve to show how all parts of the body may change, some of the changes being so slight that they would be overlooked except by an expert, others so great that the characters affected depart far from the original one.

It is important to note that the mutant genes in the X-chromosomes are not limited to any part of the body, nor do they affect more frequently a particular part. The same statement holds equally for all of the other chromosomes.

In fact, since each factor may affect visibly several parts of the body at the same time there are no grounds for expecting any special relation between a given chromosome and special regions of the body. It cannot too insistently be urged that when we say a character is the product of a particular factor we mean no more than that it is the most conspicuous effect of that factor.

If, then, as these and other results to be described point to the chromosomes as the bearers of the Mendelian factors, and if, as has been shown, these factors have a definite location in the chromosomes, it is clear that the location of the factors in the chromosomes bears no spatial relation to the architecture of the body.

Chapter 8

Sex Linked Inheritance

When we follow the history of pairs of chromosomes we find that their distribution in successive generations is paralleled by the inheritance of Mendelian characters. This is best shown in the sex chromosomes. In the female of Drosophila there are two of these chromosomes that are called *X*-chromosomes; in the male there are also two, but one differs from those of the female in its shape, and in the fact that it carries none of the ordinary genetic factors. It is called the *Y*-chromosome.

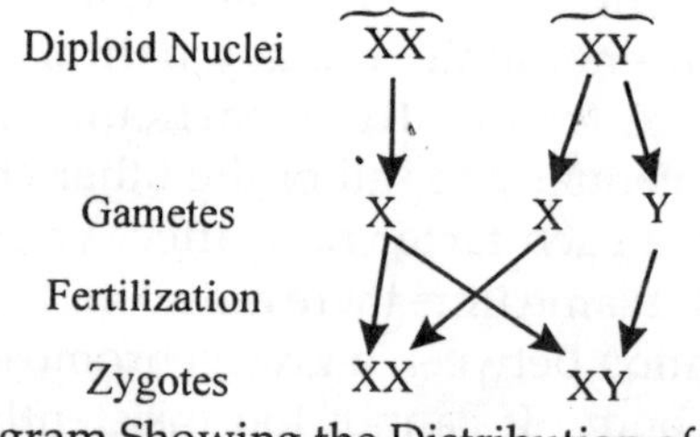

Fig. Diagram Showing the Distribution of the Sex Chromosomes from Parents to Offspring

The inheritance of a pair of characters whose genes lie in the *X*-chromosomes crosses between a white eyed and a red eyed individual. The first of these represents a cross between a white eyed male and a red eyed female.

The *X*-chromosome in the male is represented by a bar, (w), the *Y*-chromosome is bent. In the female the *X*-chromosomes are *W* and *W*. Each egg of such a female will retain one *X* (with *W*) after the polar bodies have been thrown off. In the male there are two classes of sperm—the femaleproducing, carrying *X* (with w); and the maleproducing, carrying the *Y*-chromosome. Any egg fertilized by an *X*-bearing sperm will produce a female with red eyes, because the *X*-chromosome (W) from the mother carries the dominant factor for red. Any egg fertilized by a *Y*-bearing sperm will produce a male with red eyes because he gets his X-chromosome (W) from his mother.

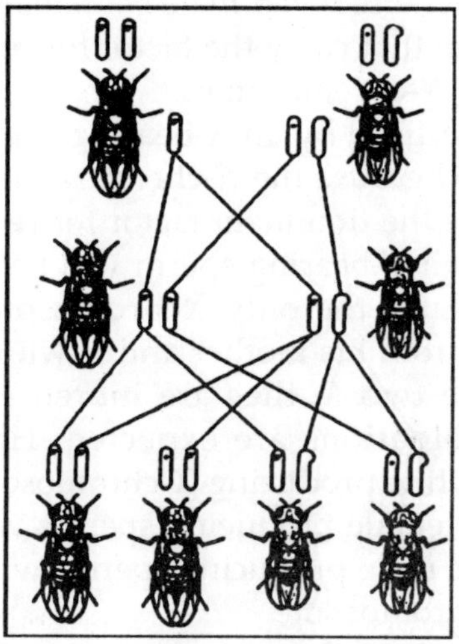

Fig. Cross of a Red eyed Female and a White eyed Male of the Vinegar fly, Showing Sex linked Inheritance

When these two F_1 flies (second row) are inbred the following combinations are expected. Each egg will contain a red eye producing *X*, (W), or a white eye producing *X*, (w), after the polar bodies have been extruded. The male will produce two kinds of sperms, of which the female producing will contain a red eye producing *X*. Since any egg may by chance be fertilized by any sperm, there will be the four classes of individuals shown in the bottom row of the diagram. All the females will have red eyes, because irrespective of the two kinds of eggs of the female, all the female producing sperm carry a (w) *X*.

Half of the males have red eyes, because half of the eggs had each a red producing *X*-chromosome. The other half of the males have white eyes, because half of the eggs had each a whiteproducing *X*-chromosome.

Evidence from other sources shows that the *Y*-chromosome of the male is indifferent, so far as these Mendelian factors are concerned.

A white eyed female is mated to a red eyed male (top row). Each of the mature eggs of such a female contains one white producing *X*-chromosome, represented by the open bar in the diagram. The redeyed male contains female producing *X*-bearing sperm, that carry the factor for red eye colour, and male producing Y-chromosomes.

Any egg fertilized by an *X*-bearing sperm will become a red eyed female because the *X*-chromosome that comes from the father carries the dominant factor for red eye colour. Any egg fertilized by a *Y*-bearing sperm will become a male with white eyes because the only *X*chromosome that the male contains comes from his mother and is white producing.

When these two F_1 flies are inbred (middle row) the following combinations are expected. Half the eggs will contain each a white producing *X*-chromosome and half a red producing. The female producing sperms will each contain a white *X* and the male producing sperms will each contain an indifferent Y -chromosome.

Chance meetings of eggs and sperms will give the four F_2 classes (bottom row).

These consist of white eyed and red eyed females and white eyed and red eyed males. The ratio here is 1:1 and not three to one (3:1) as in other Mendelian cases. But Mendel's law of segregation is not transgressed, as the preceding analysis has shown; for, the chromosomes have followed strictly the course laid down on Mendel's principle for the distribution of factors. The peculiar result in this case is due to the fact that the F_1 male gets his single factor for eye colour from his mother only, and it is contained in a body (the *X*-chromosome) that is involved in sexdetermination, while the mate of this body, the *Y*chromosome, is indifferent with regard to these factors.

In human inheritance there are characters that show this same kind of transmission. Colour blindness, or at least certain kinds of colour blindness, appears to follow the same scheme.

A colour blind father transmits through his daughters his peculiarity to half of his grandson, but to none of his granddaughters The result is the same as in the case of the white eyed male of Drosophila.

Colorblind women are rather unusual, which is expected from the method of inheritance of this character, but in the few known cases where such colour blind women have married normal husbands all the F_1 sons inherit colour blindness from the mother. Here again the result is the same as for the similar combination in Drosophila.

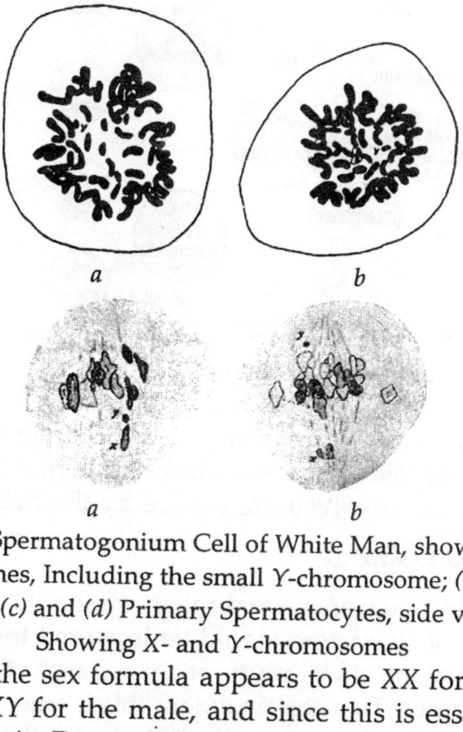

Fig. *(a)* Spermatogonium Cell of White Man, showing 48 Chromosomes, Including the small *Y*-chromosome; *(b)* Same of negro; *(c)* and *(d)* Primary Spermatocytes, side views, Showing *X*- and *Y*-chromosomes

In man the sex formula appears to be *XX* for the female and *XO* or *XY* for the male, and since this is essentially the same as that in Drosophila, the explanation of sex linked inheritance is the same. According, to de Winiwarter there are 48 chromosomes in the female and only 47 in the male, the *Y*-chromosome being absent. After the extrusion of the polar bodies there should be 24 left in the egg. In the male at one of the maturation divisions the single *X*chromosome passes to one pole. In consequence there are two classes of sperms in man; female producing containing 24 chromosomes, and male producing containing 23 chromosomes. If the factor for colour blindness is carried by the *X*-chromosome its inheritance in man works out on the same chromosome scheme and in the same way as does white eye colour (or any other sex linked character) in Drosophila, for the *O*-sperm in man would be equivalent to the *Y*-sperm in the fly.

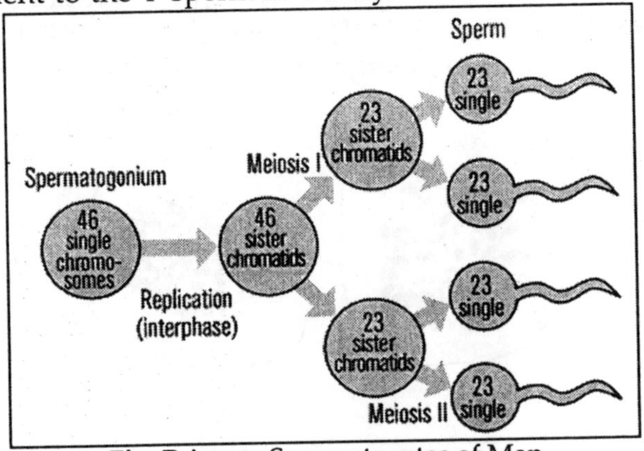

Fig. Primary Spermatocytes of Man

Painter's later evidence, showing that there is a small *Y*-chromosome in the male that acts as the mate of the *X*-chromosome, is very convincing. Whether in man the male is *XO* or *XY* the explanation of sex linked inheritance is the same since the factors involved are carried by the Xchromosomes.

CROSSING OVER

IF the linkage were never broken we should expect to find that groups of characters would be inherited together. There would be as many such groups of characters as there are pairs of chromosomes. To a certain extent this is true, but the study of the inheritance of two or more characters in the same linkage group has revealed a further fact of great interest, namely, that there takes place an interchange at times between the two members of the same linkage group, and, it may be added, only between members of the same linkage group and never between different linkage groups. This interchange gives rise to a new phenomenon in inheritance that is called crossingover, which may be illustrated by a few typical examples from Drosophila.

There are two mutant characters, black body colour and vestigial wings, whose genes lie in the second chromosome. If a fly having these two characters is crossed to a wild type fly with normal colour and long wings the offspring are like wild type flies, because normal dominates black and long wing dominates vestigial wing.

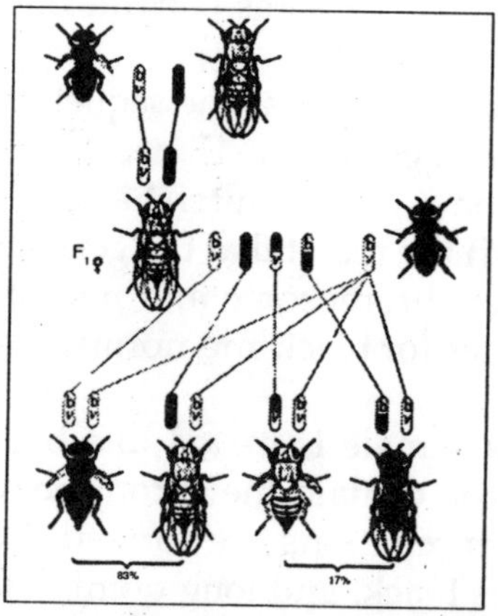

Fig. The Two Mutant Characters, Black and Vestigial, are Linked, as are their normal Allelomorphs, Gray and Long

If one of the daughters (F_1) from this cross is now mated to a male with black colour and vestigial wings (both recessive characters) the offspring are of four kinds. Two of these are like the grandparents, one black vestigial, the other normal long, but the other two kinds have, as it were, interchanged these characteristics. One kind is black long, the other normal vestigial.

These four kinds do not appear in equal numbers, but 83 per cent are like the grandparents and 17 per cent are the cross overs. The result may be stated in another way. The mutant characters that went in together, black vestigial and normal long, have remained together (linked) in 83 per cent of the grandchildren, while in 17 per cent of the grandchildren there has been an interchange or crossing over. The results may be represented in terms of chromosomes as follows. The gene for black is represented by *b* and its normal partner by *B*; the gene for vestigial by *v* and its partner by *V*. Black (*b*) and vestigial (*v*) are represented in the figure as contained in the same chromosome (here a rod) in one parent, and gray (*B*) and long wing (*V*) by the corresponding chromosome in the other parent, here by the all black chromosome.

The daughter contains one of each of these two chromosomes. When her eggs mature these two chromosomes separate and some of the eggs contain one, some the other chromosome.

These two kinds of eggs as the sequel shows represent 83 per cent of all the eggs. But in 17 per cent of the cases there has occurred in some way an interchange between these two chromosomes with the result that the genes for black and long come to be in one chromosome and normal and vestigial in the other. The genes for black and normal may be said to have crossed over.

When such a female is mated to a black vestigial male, whose spermatozoa contain the chromosome with black and vestigial, four combinations are formed. Remembering that normal dominates black, and long dominates vestigial, it will be seen that four kinds of offspring are expected. If a female with white eyes and yellow wings is crossed to a wild male with red eyes and gray wings, the sons are yellow and have white eyes and the daughters are gray and have red eyes. If two F_1 flies are mated they will produce the following classes:

Yellow White	Gray Red	Yellow Red	Gray White
98.5%		1.5%	

Not only have the two grandparental combinations reappeared, but in addition two new combinations, viz., gray white and yellow red. The two original combinations far exceed in numbers the new or exchange combinations. If we follow the history of the *X*-chromosomes we find that the larger classes of grandchildren can be explained if *X*-chromosomes are transmitted in their entirety from one generation to the next.

The smaller classes of grandchildren, the exchange combinations or cross overs, can be explained by an interchange taking place between the chromosomes in the hybrid (F_1) female. This is indicated in the diagram.

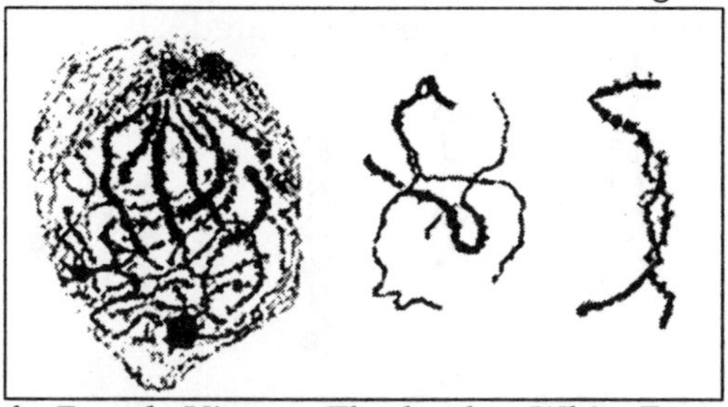

Fig. Cross of a Female Vinegar Fly that has White Eyes and Yellow Wings to a Wild type Male with Red Eyes and Gray Wings, Illustrating Crossing over

The explanation of crossing over rests on the assumption that members of the same pair of chromosomes may at times interchange. If the chromosomes are the bearers of the genes there can be no doubt from the genetic evidence that such an interchange takes place. When we turn to the known behaviour of the chromosomes in the ripening of the germ cells we find certain stages where such a process may seem possible.

At the ripening period of the germ cell the members of each pair of chromosomes come together. In several forms they have been described as meeting at one end and then progressively coming to lie side by side. At the completion of the process they appear to have united along their length.

It is always a maternal and a paternal chromosome that meet in this way and always two of the same kind. It has been observed that as the members of a pair come together they occasionally twist around each other. If where they overlap they should break and the ends unite with the corresponding ends of the opposite chromosomes, the conditions of crossing over would be fulfilled.

Unfortunately the evidence that crossing over takes place at the time of maturation, as the result of overlapping of the chromosomes, is very meagre and by no means conclusive, nevertheless, as far as it goes, this evidence is favorable for such an interpretation of genetic crossing over as that given above. From the genetic evidence for crossing over it is possible to determine the relative location of the genes in the chromosomes. The method can not be given here in detail but the general point of view may be stated.

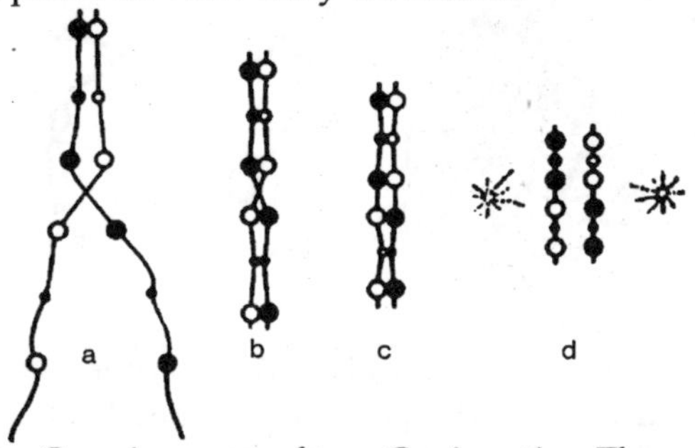

Fig. Crossing over of two Conjugating Threads

If the genes lie along the length of the chromosomes and if crossing over is as likely to occur at one level as at another, then, the nearer together two genes lie the less likely is a break between them, or conversely the further apart in the chromosome they lie the more likely is crossing over to take place. In other words the percentage of crossovers is an index of the distance apart of the genes. From such a chart one is enabled to calculate what the inheritance of any gene will be with respect to any other gene in its group provided its relation to two other genes is known.

The theory of crossing over enables the geneticist to predict the results of a given experiment with the same precision that Mendel's two laws allow prediction for a single pair of characters in the same chromosome pair, or for two or more pairs of characters in different chromosome pairs.

ETHOLOGY AND BEHAVIORGENETICS

The application of genetics to behaviour precedes by millennia the recognition of genetics as an experimental science. Vergil advised Roman herdsmen to breed their well dispositioned cattle. Plato favored the education of talented men and women in the same institutions to increase the probability of marriage between the best specimens of both sexes. Much later, Darwin recognized behaviour as well as structure as a factor in natural selection. The field we now call behaviour genetics includes practitioners with backgrounds in genetics, animal behaviour, the psychology of individual differences, the neurosciences, pharmacology, and biochemistry.

Research necessarily requires competence in disciplines other than genetics in the narrow sense. Defining and measuring behavioral phenotypes is extremely important. Ethologists have made contributions to behaviour genetics, but fewer than might be expected considering the scientific problems that are common to the two fields. My purpose is to review some of these past findings and to look toward the future.

Three major themes are prominent in behaviour genetics. The first is concerned with the importance of genetic variation within populations as a cause of individual differences in behaviour, and with the mode of transmission of these differences. These are often known respectively as heritability and genetic architecture. Both deal with processes within specified populations and are not necessarily generalizable to a species as a whole. A trait may show high heritability in a heterogeneous colony of mice and low heritability in a small closed population; a trait may appear dominant in one cross, recessive in another.

A second area of major interest is centered on biological contributions to the behavioral development of individuals and their interactions with environmental variables. Genes and genotypes can be regarded as treatments that modify the course of behavioral development. So also are physical factors ranging from nutrition to parental care. Sometimes mutant genes become figurative scalpels that excise specific biochemical processes and enable an investigator to deduce developmental patterns. Genes of this type are rare in natural populations, but are carefully preserved in fruit flies, mice, and a few other species. Numerous comparisons among inbred strains of such species demonstrate that genotypes modify behavioral development even though the species typical patterns are retained. The interplay of genotypes and the physical and social environment during development is a major interest of behaviour geneticists.

The third area, evolutionary behaviour genetics, is somewhat less developed, but is now attracting more interest. Its devotees are concerned with the molding of the gene pool of a species through behavioral adaptation to environmental opportunities and hazards. Obviously, such a process can occur only if the behaviour of interest is heritable and the population is genetically heterogeneous. Behavioral adaptation is not distinct from structural adaptation; the two are simply different aspects of a single process.

The case for the interdependent evolution of brain and behaviour has been persuasively presented by Jerison. However, the more subtle forms of behavioral variation may depend on minute structural features of the central nervous system that are difficult to observe and quantify. The behaviour geneticist's approach to evolution differs from that of most sociobiologists. The latter group work in the framework of population genetics, and their usual genetic model involves a single locus with two alleles.

Williams clearly proclaims sociobiology's disinterest in the pathways between genes and behaviour with these words: "Sociobiology does not address questions of behavioral ontogeny it is of little help in establishing links among genes, development, learning and socioculture." In contrast a behaviour geneticist places emphasis on polygenic systems and uses the genetic architecture underlying behavioral variation at different stages of development to deduce the nature of selective forces at different stages of development. Ethologists should be interested in both approaches.

GENES AND ADAPTATION

A species can adapt to a variable environment in one of three ways, two of which are genetic. It can retain a large reservoir of genetic variation, thus increasing the odds that under any stressful condition some members of the group are preadapted, and survive. In fact, most species in nature are genetically heterogeneous and the maintenance of the variability is assumed to be due to natural selection. Genetic uniformity is restricted to small, isolated, inbred populations. Another strategy is for a species to move in the direction of genetically programming a complex nervous system that modifies itself adaptively during development by responding to environmental stimuli.

The two modes of adaptation are not mutually exclusive and, in fact, often coexist in different proportions. We must be careful, however, not to infer genetic variability from behavioral variability, or vice versa. As an example, Ebert and Hyde selected for high and low aggression from a phenotypically intermediate wild population of mice. The range of phenotypes in the selected lines greatly exceeded that in the foundation population.

Underneath the apparent phenotypic similarity of the wild mice was the genetic potential for extreme variation upward and downward. On the other hand, those of us who work with inbred strains of mice, each composed of genetically identical individuals, are often frustrated by the amount of behavioral variability within a single strain. Unfortunately for many experimenters, the behavioral phenotype is not rigidly predetermined by the genotype.

These two strategies of adaptation appear to be consequences of different situations. A species living in a stable, predictable environment will prosper best with a fixed behavioral repertoire that is precoded in its genotype as a plan for a specialized, unmodifiable nervous system. A species living in an unpredictable environment will do better to have its genotype programme the development of a plastic nervous system with the capacity for modification by appropriate stimuli.

If part of the environment is predictable, a probably universal situation, selection will favour programming fixed responses for the predictable but not for the unpredictable components of its surroundings. A complete separation of these two modes of adaptation is impossible. Thus, modifiability of behaviour by experience along with genetic variation in its efficiency has been demonstrated in insects. And there appear to be biological constraints on learning even in such large brained creatures as humans.

TYPICAL BEHAVIOUR

This discussion of the relations between behaviour genetics and ethology begins with two questions. How uniform is species typical behaviour? How much of its variation is attributable to genetic differences? In general, members of a species share a common ethogram as well as a common structural plan.

Even though taxonomists depend primarily on structure as a basis for classification, field observers know that behaviour is often a reliable means of distinguishing between physically similar species. The most striking examples are sibling species, par ticularly common in insects, fish, and amphibians although a few examples are known in birds and mammals. These species, however, are not our primary concern here.

There is ample evidence of behavioral polytypy within good species as defined by the free interbreeding of their members. Such polytypy is likely to be associated with social organization or food gathering, and explanations are frequently given in terms of ecological pressures and adaptive response.

Three major hypotheses could explain intraspecific behavioral heterogeneity:

- Variant behaviour is transmitted from adults to young by learning;
- Behavioral variation is an expression of genetic heterogeneity within the species; and
- Behavioral variation arises because individuals develop differently depending on the environment they encounter.

Hypothesis 1 is in a broad sense cultural, as it implies the handing down of traditional modes of behaving from generation to generation. Cultural transmission is nongenetic except to the extent that it requires a nervous system with a capacity for learning to model the behaviour of others. In humans, cultures are a major source of behaviour variation, although their prevalence within our species does not logically exclude a genic contribution to group differences.

A well known example of a group characteristic in animals is variation in food habits among troops of Japanese macaques. In one group the habit of washing food was apparently begun by one young female and was adopted quickly by her companions. It was not found in other troops living under similar circumstances. Cultural transmission of this type may be more important in nonhumans than is generally recognized, but because it has not been widely investigated from the genetic point of view I shall not consider it further.

Hypotheses 2 and 3 emphasize individualistic responses to environmental challenges rather than modeling behaviour on the actions of others. Culture is involved only as a part of the environment in which development takes place. The difference between the hypotheses lies in the role of genetic variation. Hypothesis 2 places a premium on genetic heterogeneity as a means of ensuring that some individuals will be able to survive and propagate in environments that deviate markedly from the species optimum. This is clearly the classical concept of natural selection, and it is supported in a general way by the universality of genetic variability in natural populations.

Hypothesis 3 does not deny the role of natural selection in the past, but for the present it emphasizes the lability of phenotypic development and adaptive responses to particular environmental conditions. For adherents of this view, behavioral variation it within a species is more a matter of differences in individual life histories than in individual genotypes. Actually, the three hypotheses are not mutually exclusive. All are supported by evidence, but their relative importance varies according to circumstances.

BEHAVIORAL POLYTYPY

An extensive literature deals with variation in the behaviour of domesticated and laboratory animals, house mice, rats, dogs, fowl, and fruit flies. In these species it is often possible to separate experimentally the effects of genetic and environmental factors. Genetic variation has been demonstrated for activity level, emotional reactivity, eating and drinking, learning ability, and social behaviour. In some domestic species, selection for behaviour has produced breeds with special aptitude for certain tasks. These breed differences persist even when animals are reared as nearly as possible in the same manner.

As an example, Fuller trained five breeds of dogs to walk on a lead from their outdoor living quarters to a laboratory and found striking differences in their reactions. Shetland sheep dogs crowded against their guides, a pattern that strongly resembled the herding of sheep by their working relatives. Basenjis, with a short history of domestication, struggled more violently and bit their leash more frequently than any other breed.

Beagles, who have been selected to track small game and to give voice as they follow a trail, were noteworthy for their mournful vocalization. None of these behaviors was restricted to a single breed, and none was completely absent from any breed. Their frequencies and intensities differed enough, however, to define significantly different breed profiles. Crosses between Basenjis and cocker spaniels, breeds with widely differing profiles, yielded hybrids that were intermediate to their parents on all characteristics except vocalization, which was higher than that of either parent.

The occurence of stable, intraspecific behavioral variation among populations is ethological polytypy. It is not restricted to domesticated animals. Polytypy has been reported within single taxa whose range extends over a wide territory that covers different environmental conditions. Examples include baboons (Papio anubis), vermets (Cercopithecus aethiops), and langurs (Presbytis entellus), whose social organization varies greatly over their ranges and appears to be correlated with the nature of the habitat.

Hendrichs has recorded variations in the social organization of three species of antelope, diddik (Madoga kirki), reedbuck (Redunca redunca), and waterbuck (Kobus ellipsiprymnus), that are associated with population density, that is, in turn, presumably regulated by the supply of food and other resources.

Stacey and Bock report that acorn woodpeckers (Melanerpes formicovorus) that ordinarily live in permanent groups and defend year round territories deviated from this pattern in one location. Here, they mated in pairs and migrated in early winter. Old field mice (Peromyscus polionotus) from different geographical areas were found to have identical patterns of sexual behaviour, but these varied in quantitative aspects such as latency, frequency, and duration of the components that make up the total mating act.

There is no evidence that these polytypies are attributable to genetic heterogeneity. In fact, many of the cited authors generally ascribe these variations in social organization and behaviour to environmental pressures.

The assumption, not always explicitly stated, is that given the same circumstances any member of the species could adopt either of the alternate patterns. However, there have been few tests to disprove a genetic basis, and it would be difficult to make one with many of these species. It is possible in species that can be maintained in large numbers in captivity and bred systematically.

BEHAVIORAL POLYTYPY

Important contributions to the genetic analysis of ethological polytypy have been made by King and his students. The genus *Peromyscus* includes many taxa that have been classified at the subgeneric, species group, species, and subspecies level in ascending order of closeness of relationship. King, Price, and Webber compared animals from nine taxa on five motor behaviors: swimming, running, climbing, digging, and gnawing. They were searching for correlations between behavioral similarities and closeness of relationship as judged by taxonomists.

Secondarily, they also looked for correlations between behaviour and habitat. For example, burrowing species might be expected to excel in digging, woodland species in climbing.

In all tests the only motivation to perform any of the actions was to escape from an unfamiliar place. Differences among taxa were found on every task, but the expectation that the degree of similarity would correlate with the degree of relationship was not fulfilled. Surprisingly, the behaviour—habitat correlations were no better.

Although the results of this study were disappointing, King rethought the problem and carried out other experiments. Instead of comparing taxa on specific motor acts in an artificial setting he designed a set of tests to measure more general behavioral characteristics: speed of adaptation, reflex responses, abstractive and strategic behaviour. Again the subjects were *Peromyscus* taxa of varied degrees of relationship and similarity of habitat. The results of this study lead to a revised view of behavioral polytypy as related to genetics. All taxa of *Peromyscus,* and many other "mice," are roughly similar in form and are capable of "rodentlike behaviour" such as gnawing, grooming, and nest building.

The major differences among them lie in their modes of organization of activities in new situations as they search for food, mates, or shelter. Some taxa become specialized through fixation of genes favoring survival under typical natural conditions; others retain their juvenile plasticity into adulthood and adapt as best they can to situations as they arise. King finds evidence that both Hypothesis 2 (fixed genetic programming) and Hypothesis 3 (flexible neural programming) apply to *Peromyscus.*

In a somewhat similar experiment with wild *Mus musculus,* captured in three ecologically distinct areas and reared similarly in a laboratory, Plomin and Man osevitz found differences in open field activity, defecation, climbing, running wheel activity, and nest building. They attributed the differences to genetic factors, although controls for maternal effects and cross breeding experiments were not included in their study. A repetition with more subjects and additional controls would be useful.

In muroid rodents, male copulatory behaviour has been studied from a genetic point of view. Dewsbury (1979b) reported basically similar factor patterns in laboratory rats, the descendants of a wild population of house mice, and deer mice. In spite of the factorial similarity each of these species has a distinctive style of courtship.

In more closely related taxa, such as subspecies, male copulatory behaviour is much more uniform and for practical purposes qualitatively identical. However, significant quantitative differences in the latencies and durations of phases of mating were found in 11 of 14 measures. Subjects were three subspecies of *Peromyscus maniculatus: P.m. bairdii, P.m. blandus,* and *P.m. gambeli.* Cross fostering of deer mice on house mice produced minor quantitative effects on latencies, but, according to Dewsbury: "basic motor patterns in cross fostered mice were identical to those of normally reared animals."

A particularly interesting instance of behavioral polytypy occurs in the squirrel monkey (*Saimiri sciurus*). Two populations of this species differ markedly in the phonic characteristics of their extensive vocabulary of calls. Externally the populations are distinguishable by facial markings. The Roman arch type found in Peru, Costa Rica, and Panama, and the Gothic arch type resident in Columbia and Guyana also differ in their karyotypes.

That the dialects differ for genetic rather than cultural reasons was demonstrated by rearing infants with devocalized mothers and finding that they conversed in the dialect appropriate to their chromosomes and emitted these calls in their appropriate context. A personal communication from Newman adds that new research has shown that isolation peeps of these two forms differ so much that analysis of sonograms permits classification of individuals with 99per cent accuracy.

The recorded IPs of eight F_1 hybrids were matched with their phenotypic appearance. Five hybrids with Roman arches emitted typical Roman IPs.

The IPs of the three Gothic type hybrids were variable, one strongly Gothic, one strongly Roman, and one mixed. These observations are insufficient to hypothesize a mode of inheritance, but apparently vocalization patterns are associated to a degree with physical characteristics.

BEHAVIORAL POLYTYPY

The traditional genetic procedures of crossbreeding and selection. Most knowledge of the formal genetics of social behaviour comes from experiments with selected breeds or inbred lines of house mice, Norway rats, dogs, *Drosophila*. Can results obtained with these long domesticated animals lead to principles that can be extended to their wild relatives?

Lorenz asserted that results from animal strains that have been bred for a long time in captivity are not applicable to free living animals: "because captivity changes all hitherto effective selective factors in so profound a way that serious changes must be expected in the genome of the stock after only a few generations." Recent research indicates that Lorenz was too sweeping in this statement.

Domestic albino rats transferred from their protective laboratory environment to outdoor pens survived severe northern winters, maintained a stable population, constructed and lived in burrows indistinguishable from those of wild rats, and in general behaved like their wild cousins.

Given their freedom, the albinos were able to cope with stresses that their recent ancestors had never encountered. Of course, in a completely feral state their impaired vision and conspicuous coloration would make them more vulnerable to predators. The important point of Boice's study is that there was no evidence of a loss of species typical adaptive behaviour. Disuse of skills does not by itself lead to their deletion from the behavioral repertoire.

In a similar vein, Miller made extensive observations on Peking ducks that have been separated from their wild mallard ancestors for centuries. Pekings differ from mallards in important characteristics: They are too heavy to fly, and they lack the mallard's striking sexual dimorphism in plumage. Yet, when observed in a natural setting during three breeding seasons, they performed all the species typical displays that have been reported for mallards.

Miller concluded that his data: "offer no justification for the view that domestication has a 'degenerative' effect on social courtship behaviour patterns, but are rather supportive of the view that the customary nonoccurence of these displays is a function of the inhibitory or unfavorable environmental context in which the domestic birds usually find themselves."

The observations of Boice and Miller do not negate the existence of behavioral differences between domesticated animals and their wild progenitors but they support the idea that many such differences are the results of selection for a characteristic desired by their breeders: docility in laboratory rats, and flightlessness in birds that are intended for the dinner of their custodians.

Some changes accompanying domestication probably are the effects of the loss of dominant alleles through inbreeding. Lynch found that inbred wild house mice, compared with random bred animals from the same foundation stock, had smaller litters, bore lighter offspring, and constructed poorer nests. Random bred wild mice reared in a laboratory were not larger, more fertile, or better nest builders than standard inbred strains.

In fact, their nests were smaller than those of the heterogeneous HS stock produced by first intercrossing eight inbred strains and then continuing with a random breeding system. Lynch explains the hypertrophy of a behavioral trait related to fitness in a stock synthe sized from inbred strains that built only normal sized nests by postulating that some deleterious recessive alleles present in wild mice were eliminated during inbreeding. She believes that the inbred lines that survive the inbreeding process are good material for analyzing the genetic architecture of a species.

Her view is contrary to that of Smith who found that the heritability of shuttle avoidance learning was much lower in lines of mice selected from a wild stock than in inbred strains and heterogeneous stocks. The relevance of shuttle box avoidance to biological fitness. There is considerable evidence that good performance in the apparatus depends less on learning ability than on the prepotent response to alarm signal, freezing or running. Domestic rats have been selected for docility, and this may indeed disqualify them for inferring selection pressures in nature that are related to escape from shock.

Some of the changes in behaviour that accompany domestication seem to be the effects of inbreeding rather than of conscious or unconscious selection. Connor compared wild house mice reared in a simulated natural environment with laboratory reared wild mice and with three standard inbred strains. Wild mice, irrespective of their habitats, were more aggressive toward intruders, vocalized more when handled, and were more difficult to recapture when they were freed in a room.

In a second experiment, Connor followed behavioral changes over 10 generations in random bred wild mice maintained in a natural or a laboratory habitat. A third group from the same original population were inbred by brother sister matings. No behavioral differences were found between the laboratory and the seminaturally reared random bred stocks. Maternal effects were looked for in a cross fostering study, but none were found. In contrast, the inbred wild lines were less aggressive and easier to recapture.

The evidence is convincing that we should not ascribe intraspecific variation in behaviour exclusively to genetic or to environmental factors without evidence from experiments designed to separate their effects. Price, on the basis of his studies of the modification of activity by mode of rearing in wild and semi domesticated *Peromyscus maniculatus bairdii* came to the following conclusions. The genetic requirements for a behavioral character should not change during breeding in captivity unless:

- It is selected for deliberately; or
- It is selected for unwittingly because it improves fitness under the conditions of captivity. These conclusions are encouraging. By a judicious mixture of field observations with controlled breeding and rearing in captivity, we someday may be able to determine the extent to which natural variability in

species typical behaviour is a function of genetic differences. An example is the work of Moss and Watson, Sociobiology. Genetic heterogeneity is an absolute necessity for evolution; we know that it exists, but we do not know how important it is with respect to differences in behavioral fitness.

GENOTYPE ENVIRONMENT INTERACTION

Neither genotype nor environment is sufficient to determine a behavioral phenotype. Without a genotype there is no organism. For a zygote or parthenogenetic ovum to develop into an adult there must be an appropriate physical and biological environment. Ethologists are especially interested in the biosocial factors that shape the development of behaviour.

Indeed, the developmental process is often described as a genotype—environment (GE) interaction. The general phenomenon of joint involvement as GE coaction, and to confine GE interaction to situations in which a specific environmental change affects organisms differently depending on their genotypes. GE interactions are detected by exposing two or more genotypes to two or more conditions (e.g., stimulated versus nonstimulated) and comparing the behavioral effects.

Broadhurst and Jinks distinguished two ways in which genes might affect development:

- They could influence its stability, particularly during early stages, and effect processes that could potentially influence an individual's behaviour for life. A tendency for phenotypic development to proceed on a stable, species-specific course regardless of ambient conditions is called developmental homeostasis.
- At any stage of life, genes and their productions may influence the capacity of an organism to adjust to changes in its environment such as removal from a familiar habitat. Broadhurst and Jinks' distinction between stability and change is somewhat arbitrary,

but it is still useful. In general, stability refers to processes during early development that have generalized, permanent effects; change involves behavioral adaptations to a specific situation, with less carry over to other conditions.

GE interaction takes several forms. The most extreme is disordinal interaction where the effects of an environmental change are opposite in direction depending on the genotype of the subjects. Demonstrates that in a 2-genotype-2-treatment experiment, one could come to completely different conclusions regarding interaction depending on one's choice of treatments.

An organism must receive an appropriate exposure to some stimuli in order to develop optimally. The low end of the environmental scale corresponds to extreme stimulus deprivation; the upper end to severe stress. Strains A and B differ in their optimal stimulus requirements.

A requires less than B, but it is also more easily damaged by high levels. Suppose now that we conduct four experiments with these strains choosing only two values of stimulus intensity for each. The chosen values are indicated by x's on the abscissa.

Experiment 1 yields a clear interaction; strain A is helped by stimulation, strain B scarcely at all. Experiment 2 produces a marked disordinal interaction. Strain A is impaired as stimulation increases; strain B is improved. In Experiment 3, both strains profit from additional stimulation, and there is no interaction. Finally, in Experiment 4 both strains are impaired as stimulation increases.

Thus the nature of any interaction found may be a matter of the choice of experimental procedures rather than evidence for fundamentally different ways of responding to a stimulus change. The remedy, obviously, is to employ a wide range of treatments in any experiment where GE interactions are likely to occur.

Unfortunately, there are few experiments extensive enough to provide a test of the generality of the threshold hypothesis. They were placed separately in cages where they had no physical or visual contact with other dogs or humans. Generally the period of isolation was 12 weeks; in some experiments isolation was interrupted periodically by brief placement in an arena where subjects could interact successively with a human handler, with another puppy of the same age, and with a rubber ball or swinging toy.

From the results of a series of experiments it was concluded that the deleterious effects of early isolation are primarily attributable to an overwhelming fear reaction when an adolescent puppy suddenly faces a complex environment that is totally unfamiliar. Isolated puppies that were sedated with drugs at the time of first emergence from isolation performed almost like normally reared individuals, demonstrating that their motor and perceptual development had been little impaired by living in the small chambers.

In most mammals reactive capacities develop gradually from infancy to maturity and the environment that must be dealt with at different ages is matched with the infant's or adolescent's perceptual and motor capacities. Rather small interruptions in isolation were sufficient to counteract the effects of experiential deprivation. Puppies who had 15 minutes per week in our test arena developed social and manipulative skills almost as well as pet reared animals with several hundredfold more experience.

A number of our experiments were directed toward defining breed differences in the intensity and duration of the effects of experiential deprivation with beagles and wirehaired fox terriers as the subjects. Beagles have a history of selection for olfactory tracking of small game animals. Litters housed together in group pens seldom fought and dominance hierarchies within litters were weak.

In a familiar laboratory setting they gave the impression of relaxation, but they were somewhat shy in unfamiliar surroundings. Housed in the same way, the terriers engaged in fierce battles and formed strong dominance hierarchies within litters. When handled in tests their muscle tension was high, and they were more likely to attack than to retreat from a threat.

Thus, it was not surprising that the effects of isolation were more striking in beagles than in terriers. In one experiment, beagles, transferred to the test arena in a small cage, often failed to explore the new environment; terriers rapidly overcame an initial hesitation and reacted appropriately to objects, people, and other puppies. In another set of experiments, beagles and terriers who had been isolated from weaning to 15 weeks of age were compared with pet reared siblings. Over a period of 5 weeks each subject was observed 25 times in the arena and records made of behaviors such as responding to a handler's call, following a moving person, playing with toys, and interacting with another puppy of the same age.

A quantitative scoring system was devised for two aspects of behaviour, response intensity to key stimuli and activity index. The response indices of pet reared (control) beagles and terriers were practically identical. In both breeds, isolates were inhibited in responding during early tests. Intensity scores rose with repeated exposure to the arena; the rate of increase was much higher in terriers. At the end of the study isolated beagles were still inhibited compared with controls.

On the second measure pet reared terriers were more active than similarly treated beagles throughout the observations. During early tests, isolates of both breeds were inactive, as had been anticipated from previous experiments and, as expected, activity increased in both breeds with repeated trials. Beagles were approximately equal with their controls by the fifth week. Terriers, however, were significantly more active than their controls by the second week.

At first glance this might appear to be a disordinal interaction. Experiential deprivation makes terriers more active; it makes beagles less so. In terms of overt behaviour this is certainly true. But there are other ways of looking at the data. In both breeds isolation produced an initial aversion to the arena, reduced activity, and decreased the intensity of response to the test stimuli.

Both breeds recovered with experience, but the terriers improved more rapidly and more completely. To be sure, the isolated terriers overshot and became more active than their controls; isolated beagles were less active than their controls. In a sense, however, isolation acted similarly on both breeds in exaggerating the characteristics that differentiated them when they were raised conventionally. Normally active and investigative terriers became more so. Deliberate and cautious beagles became hesitant and timid. It appeared that isolation had made both breeds more emotionally reactive, but they expressed the change in divergent genotype typical patterns.

GENETIC ARCHITECTURE AND BEHAVIORAL PHYLOGENY

Ethology is grounded in evolutionary concepts. Roe and Simpson's *Behaviour and Evolution* brought together the views of many distinguished scientists and is still an excellent introduction to behavioral phylogeny. The value of the comparative method in deducing the phylogeny of innate behaviour patterns, particularly communication by sound, posture, and gesture, was emphasized earlier by Lorenz.

However, there are problems with the application of evolutionary theory to behaviour. The validity of the very idea of behavioral phylogeny has been criticized on the ground that it is impossible to distinguish between similarity due to common ancestry (homology) and that attributable to common selective pressures (analogy). His warning is appropriate, but it need not lead to the abandonment of the consideration of the role of behaviour in evolution.

In this presentation we are not concerned as much with homology and phylogeny as with the role of behaviour as a component of fitness and, therefore, as a directive force in evolution. Consideration of this topic requires a brief outline of theories relating the genetic architecture underlying variation in a behavioral character to the character's role in natural selection.

MENDELIAN APPROACHES TO BEHAVIOUR ADAPTATION

We can distinguish between approaches based on one or two loci (Mendelian analysis) and those based on biometrical analysis of quantitative variability. The latter type of variation is considered to depend basically on Mendelian principles, but the numerous loci that are involved are not individually identifiable. Mather called these polygenes and we now apply the term polygenic inheritance to many behavioral traits. Both the Mendelian and the biometric approaches are important to behaviour genetics.

Benzer's studies of mutants with well defined behavioral peculiarities have advanced our knowledge of neural development and function in *Drosophila*. The neurological mutants of mice have contributed to our understanding of brain development and function in mammals. Coat colour mutants, such as albinos, have been studied frequently by behaviour geneticists.

Albinos do differ from pigmented animals in many aspects of behaviour, and even normally pigmented carriers of the albino allele are impaired in some phases of development. However, neurological mutants and albinos are so rare in natural populations that they cannot be classified as true polymorphisms. We know that genetic polymorphisms are extremely common and the possibility that some of them have behavioral significance is well worth investigating.

An individual, the target of selection, is not a mosaic of characters each of which is the product of a given gene. Rather, genes are merely the units of the genetic programme that govern the complicated process of development, ultimately resulting in the phenotypic character. To consider genes as independent units is meaningless from the evolutionary point of view because the individual as a whole is the unit of selection. To regard genes as independent units is meaningless from the physiological viewpoint because genes interact with each other in producing the various components of the genotype.

For Mayr, the genotype has properties not deducible from the sum of its components. I agree. But he also acknowledges the possibility of switch genes where an allelic substitution at a single locus may shift the phenotype so that the direction and intensity of selection pressure is modified. A familiar example is industrial melanism, the replacement of light colored variants of several species of moths by darker variants in areas where pollution has altered the colour of the bark of the trees on which the moths rest.

How much evidence is there for switch genes that change behaviour drastically enough to influence evolutionary processes? Sociobiologists seem to assume that such genes are abundant enough to be important. Trivers writes: "Assume that the altruistic behaviour of an altruist is controlled by an allele (dominant or recessive) a_2, at a given locus, and that there is only one alternative allele, a_1, at that locus and that it does not lead to altruistic behaviour." Simple allelic substitution could switch such complex behaviors as altruism and selfishness or change the mating system of a species from polygamy to monogamy. It is hard to imagine how such a switch gene would operate physiologically.

Of course, any allelic substitution that affects sensory, neural, or motor functions will have some effect on behaviour, but a single mutation seems unlikely, by itself, to shift behaviour into promising new channels. However, there are reports suggesting that some genetic polymorphisms have behavioral correlates that affect selective processes indirectly.

Krebs, Gaines, Keller, Myers, and Tamarin found large fluctuations in the frequency of alleles at the transferrin (*Tf*) and leucine aminopeptidase (*LAP*) loci in the vole, *Microtus pennsylvanicus*, that were associated with changes in population density. They believe that the gene frequency changes over time were indicative of an association between the alleles at these loci and the ability to survive and propagate in stressful situations. Specifically, they hypothesized that females heterozygous at the *Tf* locus were apt to migrate during periods of increasing population density. Homozygotes, particularly for the Tf^c allele tended to remain in their original territory. The implication is that the *Tf* locus has something to do with social interactions, but, to my knowledge, this relationship has not been verified by direct observation of individual encounters. Krebs et al.'s data were obtained from field experiments supplemented by genotype determination of captured animals. A group of immunogeneticists have applied a very different approach to the genetics of social behaviour.

Andrews and Boyse claim to have mapped two loci related to mating preference in house mice. They hypothesize that one of these, *Ris* (recognition of identity signal) provides a cue, probably olfactory, by which males can distinguish preferred females. *Ris* is assigned a location in the *Qa-Tla* region of the major histocompatibility region of the mouse. Another locus, Rir (recognition of identity response) is assumed to determine whether male response is positive or negative.

The genetic procedures are complex and ingenious, but I have reservations regarding the validity of their conclusions. Males were given five to six repeated tests to determine their sexual preference. The significance of deviations from the null hypothesis (random choice) was apparently computed on the basis of the total number of tests rather than on the number of individuals.

The receptivity of the rejected females was confirmed by pairing them with other males, but one cannot exclude the possibility that factors other than identity recognition by males bias the possibility of a female being bred. In spite of these experimental problems, the hypothesis that identity recognition based on cellular characteristics is important in mating choice has evolutionary implications. More stringent statistical analysis and more adequate behavioral controls are needed.

Isolation between closely related taxa need not imply a behavioral barrier. An example that involves a physiological difference with a relatively simple genetic basis has been reported in the lacewing flies, Chrysopa cornea and C. downsei. In the laboratory these species interbreed readily. In nature, crossing is not observed. C. carnea produces three generations per summer; C. downsei produces one, then enters diapause.

Standard Mendelian crosses were made between the two species and the occurence of diapause in the hydrids was observed. The data are consistent with control of diapause by two loci. The C. downsei phenotype (diapause) requires recessive allele homozygosity at both loci. This case should probably not be called behavioral isolation in the strict sense, as the two species will mate when they are artificially manipulated. It does demonstrate that relatively simple genetic systems may have substantial evolutionary consequences.

QUANTITATIVE INHERITANCE

Let us now consider deductions that can be made regarding behavioral evolution from knowledge of the genetic architecture that regulated quantitative characteristics. It is well known that variation in such traits is largely polygenic. The theory relating genetic architecture to the evolution of behaviour has been explicated by several investigators.

Essentially, stabilizing selection for an intermediate expression of a trait, and directional selection for high or low values involve different genetic strategies. Stabilizing selection requires either additivity of gene effects, so that a heterozygote is midway between the two homozygotes; or a balance between dominant genes at different loci, some with increasing, some with decreasing action on the phenotype. Traits with such a history respond well to selection in the laboratory and are said to be highly heritable.

In contrast, selection for the extreme manifestations of a trait favors alleles with dominant effects in one direction only. Such a system ensures that the favored alleles will be effective in a single as well as in a double dose. Over time, recessive alleles will tend to disappear from the selected population, and the heritability of the trait will become low. Clearly, it is most efficient for a species to have those traits most directly related to fitness regulated by dominant genes. Falconer noted that in domestic animals heritability is lower for traits such as fertility and viability, than for traits desired by breeders for purely economic considerations.

Accepting these principles, one can work backward from determination of the genetic architecture of a trait to a reconstruction of the role of that trait in natural selection. High unidirectional dominance is evidence that extreme values of that trait increase fitness significantly. Balanced bidirectional dominance implies a history of selection for intermediate values of the trait. High additive variance indicates that a trait has not been important for fitness, or that selective pressures are inconsistent and that the ability to vary phenotypes rapidly in response to a change in environment is valuable.

There is an increasing number of studies in which the genetic architecture of a behavioral characteristic in domesticated strains has been used as the basis for inferences regarding the relation of that trait to fitness. Bruell demonstrated that F_1 hybrids between inbred strains of mice almost always surpassed both parental strains on activity tests.

The poorer performance of the parents was attributed to the inevitable, random loss of some dominant alleles during inbreeding. These losses generally involve different loci in any two strains; thus, in their hybrid offspring the deficiency of one is compensated by the contribution of the other. As might be expected, Bruell found that heterosis was more pronounced in crosses whose parents were distantly related than in crosses between lines with similar origins.

More sophisticated modes of genetic analysis were used by Broadhurst and Jinks, who measured open field activity and defecation in six lines of rats and all possible F_1 hybrids. This diallel cross design provides estimates of many genetic parameters. Their genetic findings were:

- Genes enhancing developmental phenostability tend to be dominant;
- Dominance also occurs for rapid decrease of emotional defecation on successive tests in the open field;
- Exploratory activity is also regulated by dominant genes, but these may have either a negative or a positive sign. From these observations they concluded that selection has operated to:
- Buffer the fetal and infant rat from external stress;
- Promote a rapid return to normal emotional status after arousal;

- Strike a balance between bold curiosity that could lead to disaster, and timidity that could cause exclusion from essential resources. None of this is surprising, but it is satisfying to find that the genetic findings are consonant with our expectations of what organisms should do in order to prosper.

The genetic approach to reconstruction of evolutionary history is gaining ground, but there is difference of opinion regarding the best way to proceed. For example, Lynch concluded that: "because inbred mice are not weakling, analysis of crosses between them may represent a more efficient method for detecting genetic architecture than inbreeding of wild populations." Smith disagrees on the basis of his comparison of the heritability of shuttle avoidance in selected wild mice and in inbred strains.

He doubts that it is safe to deduce the selection history of a hypothetical parent population from the genetic architecture of contemporary inbred strains of largely unknown origin. Obviously, more research is needed to settle this matter.

Henderson made an effort to escape the bind of first accepting the basic hypothesis of the relation between genetic dominance and fitness of a trait, then conducting a genetic experiment and, from its results deducing the fitness value of that trait. He deduced from observation that a particular trait was related to fitness and then conducted a breeding experiment to determine its genetic architecture.

Specifically, he reasoned that mice should be selected for low levels of motor activity during infancy because they are safest when they remain in their nest. If they are accidentally dislodged their chances of being retrieved are greater if they remain quiet and await their mother. He predicted that, in direct contradiction to the situation in adults, low activity would be dominant in infant mice.

A triple test cross confirmed his prediction; the heritability of activity was low and inactivity was dominant. This was attributed by Henderson to a relaxation of selection for infant immobility over many generations of laboratory rearing.

In a very different species, Cheng, Shoffner, Phillips, and Shapiro studied imprinting to visual and auditory stimuli in five groups of mallards (Anas platyrhyncos) ranging from a long established game farm line to 25%, 50%, 75%, and 100% wild ancestry. Wild ducklings imprinted better than domesticated ducklings, and heterosis was found in the hybrids, as would be predicted from the hypothesis that traits with adaptive value show little additive variation. Efficient imprinting seemed to be correlated with differences in arousal during early exposure to stimulation.

Confirmations of a prediction do not prove the universal value of a method. However the approach of Henderson and Chang et al. is promising and has considerable potential. As Henderson states, the method is most powerful when predictions can be made that a particular kind of behaviour will differ in its contribution to fitness at different parts of its life-span, or in dissimilar environ ments.

The method is obviously limited to species in which multiple distinct lines or strains are available. Principles derived from such studies are, however, likely to have general application.

It is only fair to state that the usefulness of this approach to defining adaptive behaviour has been challenged on the basis that the experimental subjects available for analysis of genetic architecture are not representative of natural populations, and that a history of inbreeding distorts the genetic structure typical of a species under natural selection. These warnings must be kept in mind but much of the data obtained in this way seems to substantiate accepted ideas of the characteristics of fitness enhancing behaviour. There is, of course, a possible circularity in such experiments.

One can hypothesize that high levels of behaviour B_1 are adaptive and thus should have increased fitness during recent evolutionary history. Suppose now that an experiment is performed and low levels of B_1 are found to be inherited in dominant fashion. One could conclude that the original hypothesis was wrong, and that actually low levels of B_1 enhance fitness.

Alternatively one might consider the possibility that B_1 has little to do with fitness but, in the population studied, it is correlated genetically with another behaviour, B_2, that has been selected on the basis of enhancing fitness. The best way out of such circularity is good ethological studies relating B_1 and B_2 independently to survival and propagation.

Another problem with the attempt to deduce the selective importance of a particular form of behaviour is the choice of the method of measuring it. Behaviour geneticists have been sure for years that in an open field C_{57}BL mice were more active than BALB/c mice. Whitford and Sipf found no important differences between these strains when the height of their open field was reduced from 36 inches to 1 inch by insertion of a plexiglass sheet. It is clear that open top, open field activity is very different from contact top, open field activity.

Chapter 9

Human Blood Groups

A Person who has had the measles once is extremely unlikely to get a second attack of the same disease; he is, as we say, immune to it. Many other infectious diseases confer immunity upon recovery, although not always for such a long period. Immunity is the result of a defence mechanism of the body against the invading microorganism.

Antibodies are formed which combine with the microorganisms and often destroy them by, for example, clumping or dissolving them. Lasting immunity is found when the antibodies outlast the infection.

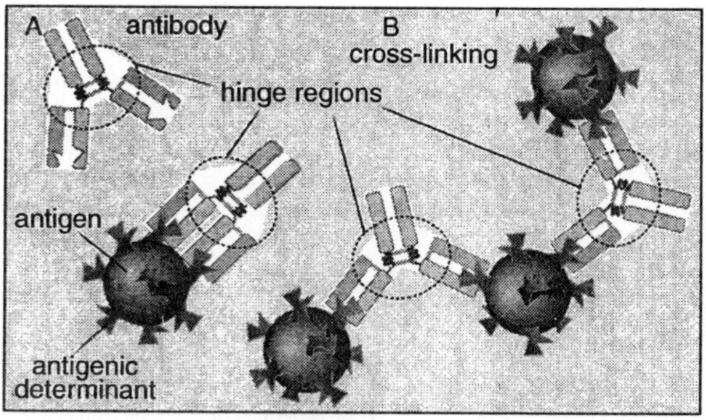

Fig. Antigen and antibody

Antibodies are highly specific in their reactions. Measles are a virus, but a person who has acquired immunity against measles is not, for that reason, less likely to contract German measles or poliomyelitis or any other virus disease. The specificity of antibodies is a consequence of the manner in which they arise. When a microorganism gets inside a warm blooded animal, its protein acts as an antigen, that is, it stimulates the production of antibodies by the infected individual.

Proteins are highly complex molecules; they consist of thousands of atoms which are arranged into specific configurations. An antigen functions like a mold for shaping a host protein into its own mirror image. When the antibody has become free of this mold it still "fits" the exact type of antigen by which it was produced, but no others.

Not only the protein of microorganisms but any foreign protein acts as antigen. This is the reason why no successful skin grafts can be made between different persons, not even between mother and child. No two individuals have exactly the same proteins in their skin cells, and when a graft is made it is eventually destroyed by the very antibodies whose formation it induces.

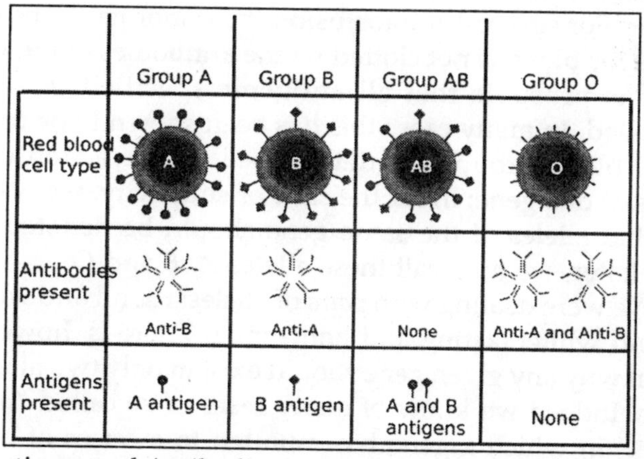

Fig. Antigens and Antibodies in the A-B-O Human Blood Groups.

Antigens are also carried by the protein of the human blood cells. Fortunately, individual specificity is here less pronounced than in skin proteins; otherwise no transfusion of a patient with whole blood would be possible.

Yet not every combination between donor and patient is compatible, and in the early days of blood transfusion fatal accidents arose from clots of transfused blood in the veins of a patient. The reason for this is now well understood, and such accidents are avoided by the right choice of donor. A number of different antigens are carried by human red blood cells. Most of them can be detected only by the fact that they induce antibodies when the blood is injected into a warm blooded animal, for example, a rabbit. But there is one group of antigens for which ready made antibodies occur in human serum, the fluid in which the blood cells are suspended. The main antigens of this group are called *A* and *B*, and human beings can be divided into four groups according to whether their red blood cells carry one of these antigens (*A* or *B*), both (*AB*), or neither (*O*).

In addition, the serum of each individual contains preformed antibodies to those antigens that are not present in the blood cells. Thus, in blood group *A* the serum contains anti-B, in blood group *B* it contains anti-*A*, in blood group *O* it contains anti-*A* and anti-*B*. and in blood group *AB* it contains neither. For successful transfusion the donor has to be chosen so that his blood is not clotted by the antibodies of the patient.

It is probable that all antigenic specificities are gene controlled. In many cases this has been shown to be true. The *A-B-O* blood groups are determined by the action of three alleles of one gene; if, for the sake of simplicity, we waive the rule that alleles of the same gene should be denoted by the same letter, we may call these alleles *A*, *B* and *O*.

We were dealing with *pairs* of alleles: round and wrinkled, red and white, tasting and nontasting. There is, however, no reason why any given gene should exist in only two alternative forms. Indeed we know of many cases of so called "multiple alleles" in which a gene has mutated to a series of different alleles.

It shows a wild rabbit and three mutant varieties, produced by the action of multiple alleles of a gene controlling coat colour. The chinchilla rabbit has silvery fur; the Himalayan rabbit is white apart from black ears, feet, nose and tip of tail, and it has pink eyes; the albino rabbit is wholly white with pink eyes. The alleles producing these effects are usually denoted by the letters C for full (wild) colour; c^{ch} for chinchilla; c^{H} for Himalayan; and *c* for albino. Any individual can, of course, carry only two of these genes and can be either homozygous for one of them, like the rabbits in the diagram, or heterozygous for two, for example, $c^{H}c$.

In heterozygotes, coat colour depends on the dominance relationship between the two alleles; since c^{H} is dominant over c, $c^{H}c$ rabbits look Himalayan.

Fig. Multiple Alleles in the Rabbit

The possible genotypes for A, B, and O in man and the blood groups that correspond to them are set out in the following table:

	homozygous			*heterozygous*		
Genotype	OO	AA	BB	AO	BO	AB
Blood group	O	A	B	A	B	AB

You will notice that each gene produces its characteristic antigen independent of its partner gene. As a result O, which does not produce an antigen, cannot be recognized in the presence of either A or B. In other words, O is recessive to A and B, and persons of blood group A or B can be of two different genotypes according to whether or not they are heterozygous for O. This plays an important role in the medicolegal application of the A-B-O blood groups.

Newborn infants in maternity wards are labeled with the

name of the mother. Very occasionally doubt may arise as to whether a mother has been given the right baby, and blood typing may then help to decide this question. In a well known case a woman, let us call her Mrs. Smith, found on her return from the hospital that her baby carried the label "Brown." There had been a Mrs. Brown in the same ward as Mrs. Smith, and she had, indeed, been given an infant labeled "Smith."

A glance at the table will show you that between them they carried nothing but O genes and could not have a child with any blood group but O. On the other hand, the Smiths could *not* have a child of blood group O; for, while Mrs. Smith belonged to blood group O, Mr. Smith belonged to AB. He was, therefore heterozygous for A and B, and had to transmit one of these alleles to his child. When it was found that baby "Brown" was O and Baby "Smith" was A, it was evident that the labels were correct and that the two women had been given the wrong infants.

Not all cases can be decided so easily and unambiguously. Suppose that both Smith and Brown parents had belonged to blood group A. Then either couple might have produced baby "Smith" with blood group A. Moreover, either couple might have produced baby "Brown" with blood group O; for A individuals may carry O genes and can transmit them to their children. Fortunately, many more blood groups, controlled by different series of alleles, have been discovered in recent years and can be used for deciding difficult cases.

This refers also to the use of blood typing in cases of disputed paternity, many of which cannot be decided on the basis of the A-B-O system alone. Mr. Smith, as we just saw, could not have fathered a child of blood group O; similarly, Mr. Brown could not have fathered a child belonging to blood group A, B, or AB. But the father of an A baby from an A mother might belong to any of the four major blood groups. In such a dilemma, recourse to other blood groups may help narrow the issue. In any case, blood group tests in cases of disputed paternity fulfill only a negative function; they can exclude certain men from possible fatherhood, but they cannot indict any particular man.

Suppose blood group O were alleged to be the father of an illegitimate child of blood group O by a woman of blood group B. On purely genetical grounds, there is no objection to this assumption, but neither is there an objection to any other man of blood group O or against any men of blood groups A and B.

Only where the baby shares a very rare blood group with the alleged father, but not with the mother, will the geneticist feel reasonably convinced of the correctness of the allegation, although the court may not accept his judgment. In 1921 a Norwegian judge made medicolegal history by convicting a man of paternity on account of a rare dominant abnormality foreshortened fingers, or "brachydactyly" which was absent in the mother but present in both the infant and the alleged father.

DANGER TO INFANTS

Like the alleles which determine ability or inability to taste PTC, the A, B and O blood group alleles occur not only in human beings but also in the large anthropoid apes. A new and highly important human blood group was discovered in 1940 through tests on lower monkeys. When rabbits were injected with blood from a Rhesus monkey, their serum was found to contain an antibody which combined with the red blood cells of about 85 per cent of individuals in a White population; the remaining 15 per cent gave no such reaction.

Persons whose blood reacts with the antibody against Rhesus blood are called Rh-positive; those who do not react are called Rh-negative. Both types of individual occur in all human races, but the proportions of positives and negatives differ. Evidently the red blood cells of Rh-positive individuals carry an antigen which is so similar to the antigen on the Rhesus blood cells that the antibody does not distinguish between them. The blood cells of Rh-negative persons lacks this antigen.

When an Rh-negative person is given a blood transfusion from an Rh-positive donor, his body reacts like that of a rabbit into which Rhesus antigen is injected: it makes antibodies against the foreign antigen. Since it takes some time before a sufficient amount of antibody is produced, a first transfusion of this kind is usually perfectly harmless; but the antibodies persist in the serum of the transfused person, and should he receive a second transfusion with Rh-positive blood there may be a severe reaction.

This explains why accidents had happened occasionally in blood transfusions between individuals of the same A-B-O blood group. No antibodies are formed by an Rh-positive individual who is given a transfusion with Rh-positive blood: for such a patient the Rh-antigen is not a foreign antigen.

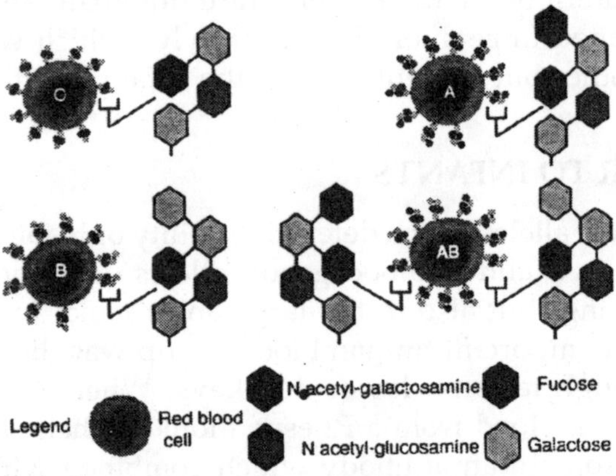

Fig. Blood Group System

Of much greater importance is the fact that the new blood group furnished the explanation for a severe and hitherto rather mysterious disease, called "hemolytic disease of the newborn." Before we go into this, we shall have to consider the inheritance of the Rh character. At first sight it is very simple. Rh-positive children invariably have at least one Rh-positive parent.

Rh-negative children, on the other hand, may issue from any combination of parents: both Rh-positive, or one Rh-positive, the other Rh-negative. Thus Rh-positive parents may carry the Rh-negative gene and transmit it to their children, while Rh-negative parents carry only Rhnegative genes. Expressed in Mendelian terms, this means that we are dealing with a dominant allele, Rh, and a recessive allele, rh. Rh-negative individuals must be homozygous (rh rh), but Rhpositive individuals may be homozygous (Rh Rh) or heterozygous (Rh rh). In reality, more alleles are known and the situation is far from simple; but for most practical purposes these complications can be disregarded.

With this simple genetical picture in mind, let us return to hemolytic disease of the newborn. About once in several hundred pregnancies among White women, a newborn child suffers from a severe anemia with jaundice which in the worst cases may be fatal if blood transfusion cannot be carried out in time.

The disease has a tendency to affect several children in the same family, and this seemed to point to a hereditary basis; but there are also special features which do not fit into any simple Mendelian explanation.

Thus first born children hardly ever are affected; but once a coùple have had one hemolytic child, more of their children are likely to be affected. In some families all children following the birth of the first affected one suffer from the disease. The explanation came when it was found that in almost all cases of hemolytic disease the mother was Rh-negative and the child Rh-positive and that, moreover, the blood of the mother contained anti-Rh antibody.

On the contrary, no such antibody is found in the blood of Rh-negative women who never have been pregnant or who have given birth only to Rh-negative children. This shows that an Rh-negative woman reacts to an Rh-positive fetus as she would to transfusion with Rhpositive blood: she makes antibodies against the Rh-antigen. As after blood transfusion, the level of antibodies takes time to build up and usually no reaction occurs with the fetus that stimulated antibody production. This explains why first born children rarely suffer from hemolytic disease

But when an Rh-negative woman is pregnant with a second Rh-positive child her antibodies may get into the circulation of the fetus and destroy its blood cèlls. When an Rh-negative woman bears an Rh-positive child, the *Rh* gene in the fetus must have come from the father's side. The father, therefore is either homozygous, *Rh Rh*, or heterozygous, *Rh rh*. In the former case, he will transit an *Rh* gene to every child; this is the type of marriage where, once a sufficiently high level of antibodies has been established in the mother, every child will be affected. In the second case, half the children will inherit an *rh* gene from both father and mother and will be Rh-negative; these children are in no danger from the antibodies in the mother's blood.

Fortunately, a large proportion of pregnancies which, from the blood groups of father and mother, might lead to hemolytic disease of the newborn do not, in fact, do so, partly because of the low level of antibodies in early pregnancies, partly because of differences in the ease with which antigens and antibodies are exchanged between mother and child, partly perhaps for still other reasons.

All the same, many hospitals make sure that pregnancies of an Rhnegative mother and an Rh-positive father get special attention. In severe cases, an affected child may be saved by the replacement of all its blood with blood that is free from antibodies.

Similar gene controlled incompatibilities between the blood groups of mother and offspring are known in a number of animal species, for example, rabbits, dogs, and horses. Not always, however, do the antibodies harm the fetus before birth. In the horses, for example, antibodies are produced when a stallion, carrying on its blood cells an antigen *R* sires a foal with this antigen by a mare lacking it. The mare produces antibodies against *R*, but it takes at least three pregnancies by an *R* stallion before the danger limit has been reached. In a subsequent pregnancy by an *R* stallion, the antibodies in the mother's milk may kill the foal; but the foal may be saved by fostering it out to another mare.

GENES AND CHARACTERS

There is hardly any character in any organism that is not to some extent under the influence of the genotype. Genes control the colors and shapes of animals and plants, the ability of animals to see, hear and smell, the resistance of a man to an infecting bacterium, and the ability of the bacterium to infect the man. They influence the rate at which a calf will grow on a given diet, the degree to which a child can profit from education, and the likelihood that a person's balance of mind will give way under mental stress.

They determine whether or not a Chlamydomonas has flagellae for swimming about, and whether or not a yeast culture can grow on galactose. Genes influence the most superficial as well as the most fundamental processes of life; they act at every stage of development. Indeed, the very chance of survival at any period from conception to extreme old age is to some degree under the control of the genotype. In chickens, rats and mice, mutant genes have been shown to kill the embryo by severely disturbing the development of such vital organs as the skeleton, the brain, the kidneys.

Doubtless, the normal alleles of these genes play a role in controlling the normal development of these organs. Equally doubtless, similar genes will play this role in the human embryo. At the other end of the life span we find that inbred lines of mice differ in the average length of life of the individual. There is little reason to doubt that also in man longevity is in part genedetermined, although the very great genetical and environmental variability among human individuals makes it difficult to provide conclusive evidence for it.

It seems that the great majority of genes are indispensable for the survival of the organism. In experimental animals or plants, very small pieces of chromosome can be destroyed by, for instance, X rays. Attempts to breed individuals that lack the same piece in both partner chromosomes rarely are successful; for no matter which piece is missing, it almost invariably seems to contain genes without which survival is impossible.

In Drosophila, it could be shown that even genes controlling such a seemingly irrelevant character as eye colour or shape of wing margin need to be present in at least one chromosome for the embryo to develop into a living fly. If such genes had no other effects than their easily observed ones, this would be difficult to understand. There is, however, ample evidence that most, if not all, genes produce a variety of effects.

This evidence comes from a study of mutant genes. The very existence of a normal gene can be inferred only after it has mutated at least once, and its effects must be deduced from a comparison between normal and mutated individuals. Many mutant genes have several obvious effects. In Drosophila, the mutant gene "lozenge" makes the eyes look sticky like boiled sweets, removes the claws and pads from the feet so that the flies can no longer walk up the glass wall of their culture bottle, and makes the females infertile.

Japanese waltzing mice are deaf; Frizzle hens have immature ovaries and lay few eggs. It is safe to conclude that the normal allelomorphs of these genes are necessary for the normal development of eyes, feet, and genetical organs in Drosophila; of ears and nervous system in mice; of plumage and ovaries in fowl. Whenever a mutant gene is studied thoroughly, it is found to produce a whole array of effects.

Most mutant genes reduce viability; it is therefore not surprising that their normal alleles are necessary for life. The ability of a gene to produce several distinct effects is called "pleiotropy," that is, "many directedness." From what has been said just now it is clear that pleiotropy is the rule and not the exception.

Just as every gene influences many characters, so is every character influenced by many genes; this is known as "gene interaction." In a way, gene interaction and pleiotropy are complementary to each other; for to state that many genes act pleiotropically on viability is only a different way of saying that viability is influenced by many genes.

Similarly, many mutant genes and, by inference, their normal alleles affect fertility and, conversely, fertility depends on the action of all these genes as well as on that of others which do not advertise their presence by striking effects on superficial characters. Even very specific characters, like the colour of the iris in the human eye, depend on the action of many genes.

In order to avoid details, the "characters" have been chosen as broadly as possible and might better be called "organs" or "functions." The choice of more narrowly defined characters, such as eye colour or structure of the tail, would have yielded an essentially similar picture.

A line connecting a gene with a character signifies that the gene influences the character. The diagram has been kept simple by restricting it to a few genes and characters. Had we entered more characters on the right, more lines would have started in most genes on the left. Had we entered more genes on the left, more lines would have come together in most characters on the right. The irregular way in which the lines cross each other shows that the constellation of characters controlled by a single gene varies between genes.

Thus the genes *fi*, *st*, *Va*, and *v* all cause a similar nervous disorder resulting in circling and shaking movements; but fidget mice can hear, while the other three mutant types are deaf; only the gene *Va* produces an irregular pattern on the coat, only *fi* leads to sores in the eyes, and only *st* shortens the tail. Again, it should be remembered that these interactions between mutant genes and characters indicate some of the ways in which the normal alleles interact in normal development.

PLEIOTROPY: ONE GENE AFFECTS SEVERAL CHARACTERS

Sickle Cell Anemia

In 1910 a young West Indian Negro consulted an American physician about a feverish cold. Since he complained of not having felt well for quite a long time, the doctor undertook a thorough examination, in the course of which he found that the young man suffered from severe anemia of a hitherto not recorded type.

Under the microscope his blood presented a striking picture. While in normal blood the red cells are uniformly round disks, the blood of the patient contained a high proportion of curiously sickle shaped red cells. Several years later the same abnormality was found in the blood of a father and his son, both suffering from anemia. The condition was called "sickle cell anemia"; it is quite distinct from the more usual forms of anemia.

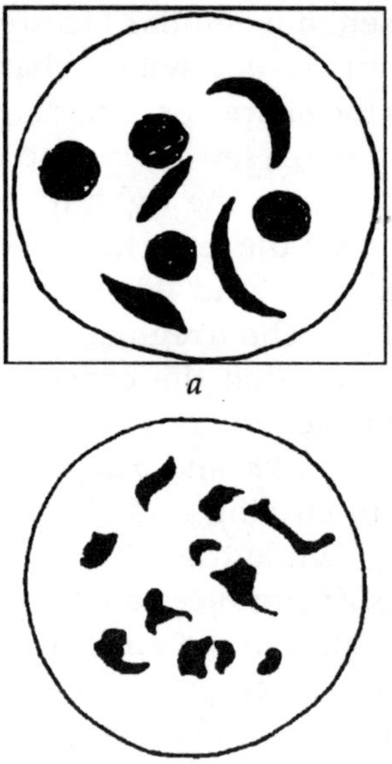

a

b

Fig. (*a*) Blood Corpuscles of a Patient with Sickle cell Anemia. In (*b*) the Sickling has been Emphasized by Keeping the Drop of Blood Sealed off from the Air

Once attention had been directed to it, more cases were reported, at first exclusively from Negroes. It soon became obvious that the disease was in some way hereditary, but its accurate genet ical basis was determined only when a technique had been developed that ensured reliable detection of the sickling phenomenon.

This technique consists in keeping the sample of blood for a day or more sealed off from the air, when the cells of patients assume bizarre shapes. With this technique it was established that, in addition to the typical blood picture found in patients suffering from sickle cell anemia, a mild, but distinctly recognizable form of sickling occurs in the blood of about 9% American Negroes who do not suffer from anemia.

This milder, harmless form of sickling is especially frequent among the relatives of patients suffering from sickle cell anemia; in particular, it is practically always found in both parents of an anemic child. This led to the interpretation, fully borne out by further investigations, that the condition is due to a gene *S*, which in heterozygotes *S*+ causes the mild form of sickling, unaccompanied by anemia, while homozygotes *SS* are anemic.

Laboratory investigations of the abnormal blood revealed that the gene *S* affects the hemoglobin, that is, the pigmented, ironcontaining protein of the red blood cells. In an electric field like the one the hemoglobin molecules of *SS* individuals move more slowly than the hemoglobin molecules of normal individuals. The blood of heterozygotes *S*+ contains both kinds of hemoglobin in approximately equal proportions; apparently one kind is formed under the influence of the *S* gene, the other under the influence of its normal allele.

This, incidentally, is a good illustration of the fact that the terms "dominant" and "recessive" are relative, and that a gene may be dominant at one level of observation and recessive at another. At the level of hemoglobin chemistry, the gene *S* and its normal allelomorph act independently of each other, similar to the blood group genes *A* and *B* at the level of the microscopic blood picture, *S* is incompletely dominant over its normal allele; finally, at the level of the manifest disease, *S* is recessive to its normal allele.

A search for the presence of the gene *S* in many countries gave curious results.

At first it seemed as though the gene was restricted to Negroes, in Africa and in the two Americas. Later, individuals whose blood could be made to sickle were found also among Italians, Greeks, and other peoples around the Mediterranean, and among South Indian tribes.

The distribution of such individuals, however, was exceedingly patchy. While large numbers of them occurred in certain parts of Africa, they were practically absent from others. Even within the same territory there often were striking differences in the proportion of individuals with sickle cells.

In Uganda, for example, this proportion was nearly one half in one tribe and nearly zero in another. The most puzzling aspect of this survey was that sicklers (S^+) have such high frequencies in certain populations. Homozygotes for *S* rarely reach adulthood and have children; nature effectively sterilizes them. Although, sterilization of homozygotes for a recessive gene is a very inefficient way of getting rid of the gene, it will keep its frequency low.

In fact, other genes that result in "natural sterilization" of the homozygotes, such as the gene for amauritic idiocy, are rare even in heterozygotes. It therefore seemed that in certain populations the childless death of the homozygotes was overbalanced by the fact that the heterozygotes had some natural advantage which gave them a better chance than the rest of the population to have children, half of whom would inherit the gene *S*.

A possible clue to the nature of this advantage was provided by the discovery that sicklers are frequent in malaria ridden regions and are rare or absent in malaria free regions. This suggested that the gene *S* in some way protects its carriers against the malaria parasite. Investigations to test this idea lent support to it.

Out of 288 young children in a malaria district of Africa, 43 were heterozygous for *S* and of these 12, that is, 28%, were found to carry the malaria parasite in their blood; of the remaining 245 children, all without the gene *S*, 113, that is, 46 per cent, carried the parasite. In one African tribe 30 volunteers offered themselves for infection with malaria; 15 of them were S^+ and of these only 2 contracted the disease; the other 15 were ++, and of these only 1 remained free of it.

Altogether, there is good, although not yet wholly conclusive, evidence for a pleiotropic effect of the gene *S*. The same altered hemoglobin that in homozygotes results in anemia makes the blood of heterozygotes an unsuitable environment for the malaria parasite. Since malaria affects physical and mental development in various ways, the final piciotropic effects of a gene controlling susceptibility to this disease will be correspondingly far reaching.

If this explanation of the curious distribution of the gene *S* is correct, we are dealing with a case in which a heterozygote for two alleles (S^+) is superior to the homozygotes for either (*SS* or ++). This situation has been mentioned as a possible contributory cause of hybrid vigour.

Hereditary Dwarfism in the Mouse

They did not breed, but when their littermates were bred together, some of the litters contained about one quarter dwarf mice. The character thus was attributed to the action of a recessive gene. Mice homozygous for this gene stop growing at the end of the second week or a little later and usually reach a weight of only 6-8 grams as compared with a weight of 20 grams or more for normal mice. An adult dwarf mouse together with its normal littermate.

The dwarf gene has many pleiotropic effects. Early on in development, even before growth has stopped, homozygotes can be distinguished from their normal littermates by blunt snouts and relatively shorter ears and tails. At the stage when normal young mice are lively and easily excited, dwarf mice are timid and quiet.

They are very sensitive to temperature changes; but when they are kept at the right temperature they tolerate starvation better than normal mice. They have a shortened life span and are sterile in both sexes.

Examination of their inner organs revealed that most of their hormone producing glands, such as the thyroid, the pituitary, and the gonads, were undersized and abnormal in structure. Now, it is well known that the pituitary, a small gland at the base of the skull, occupies a key position in the control of hormone production; for this reason it is often referred to as "the master gland." It therefore occurred to some scientists to attempt to cure the hereditary dwarfism by treatment with pituitary.

Pieces of fresh rat pituitary were implanted daily under the skin of dwarf mice which had stopped growing. The success of the treatment was spectacular. The treated animals attained normal weight and body proportions, they became lively and vigorous and behaved in every way like normal mice. The males became fertile, but the females remained sterile, although their ovaries regained normal size and structure.

Similarly, all other hormone producing glands, with one notable exception, became normal. The excep tional gland, which did not respond to the treatment, was the pituitary. This is exactly what one would expect if the basic defect, from which all the others are derived, is an abnormality of the master gland; for, while implanted pituitary can supply all the hormones by which a normal pituitary controls its subordinate glands, it cannot cure the source of the trouble in the pituitary itself.

GENE INTERACTION: ONE CHARACTER DEPENDS ON MANY GENES

Chlorophyll Formation

In the cells of green plants the green pigments, together with some yellow ones, are carried inside numerous small bodies, the so called "chloroplasts".

Primitive plants, for instance, algae, may have one large chloroplast instead. Before the invention of the electron microscope, not much detail could be discerned in a chloroplast. Nowadays electron micrographs reveal a highly complicated structure.

Combined with genetical studies observations with the electron microscope have become an important tool for understanding the role of the genes in the development of the chloroplast, and although this work is still at its beginning, a rough picture of gene interaction has started to emerge. Electron microscopic studies have shown that the normal chloroplast (V) develops by stages II-IV from a small colorless granule (I), which in the course of development not only changes its size and shape but also forms a complex interior network of lamellae between which some globules are suspended.

At certain stages of this structural process, the chloroplast forms pigments, first yellow and then green ones. In the diagram, the source of the pigments seems to lie outside the chloro0plast; this has been done for the sake of clarity: in reality, the pigments form inside the chloroplast. It is certain that a vast number of genes are involved in the control of these processes, for many different mutations have been shown to interfere with chlorophyll and chloroplast formation.

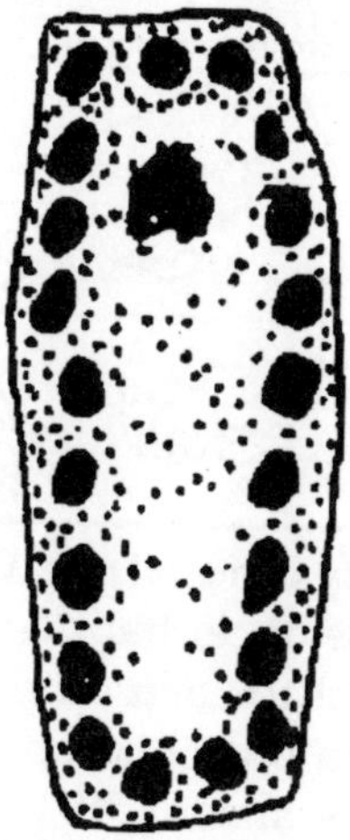

Fig. Plant Cell with Nucleus and Chloroplasts

More than a hundred such genes have proclaimed their existence by mutation; many more may have remained undetected because they have not yet given rise to observed mutations. It is also known that different genes control different steps of chlorophyll formation; for mutations may arrést chloroplast development at different stages and in different ways.

Genes Acting in Series: an Illustration from Human Life

Genes, like *A, B, D, G* or C_1, C_2, C_3 or F_1, F_2, F_3, that act in series resemble workers at a conveyer belt; nonfunctioning genes of such a series resemble absentee workers. Let us see what happens in a factory production line when one of the workers is absent. The assembly of a doll by four girls working at a conveyer belt. Ann has a supply of dolls' heads and fastens them to the bodies which are sent to her on the belt. She sends her product to Beryl, who adds the arms. Beryl sends her product to Cathie, who adds the legs. Cathie sends her product to Dot, who dresses the doll and puts the finished product into a box. Let us suppose that no worker can carry out any other manipulation than her own and, moreover, that she can do this only when the previous manipulations have been completed.

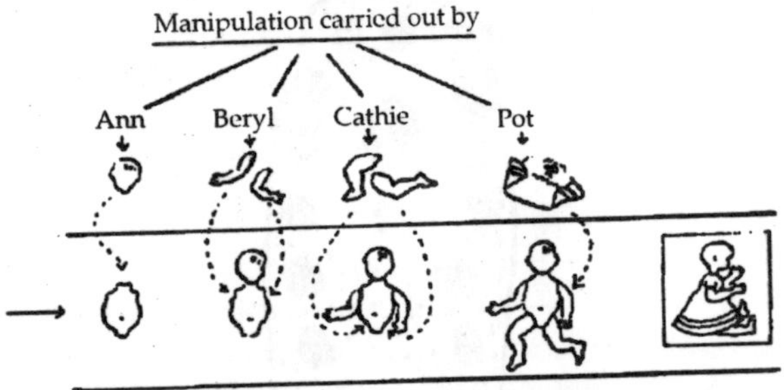

Fig. Assembly of a Doll on a Conveyor belt

This might, for instance, happen if the various parts of the body were attached to each other by strings with hooks which could be used in only one order. Now consider what happens when Ann is absent. If the conveyer belt is not stopped, headless and limbless bodies pass Ann's place, and since they cannot be manipulated by the other workers they accumulate at the end. If, instead, Beryl is absent, Ann will be able to carry out her manipulation but Cathie and Dot will not, and limbless bodies with heads will accumulate at the end of the line.

Cathie's absence will result in an accumulation of legless bodies, and Dot's absence in an accumulation of naked but otherwise complete dolls. In general, when a worker is absent there will be an accumulation of the unfinished product furnished by the preceding worker.

Suppose now that the foreman of the production line is anxious to fulfill his quota of finished dolls and that he has the chance of obtaining either the finished product or the intermediate ones from a parallel production line which has a surplus of all of them. The finished product will, of course, help him in any case; but in order to keep the remainder of the workers occupied he may prefer to ask for an unfinished one. When Cathie is absent, Dot can be occupied by a supply of naked dolls, but Ann and Beryl will have to remain idle. When Beryl is absent, Dot can again be occupied by a supply of naked dolls; but now it would be preferable to obtain the previous product, legless dolls, for this would keep both Cathie and Dot busy.

Finally, when the absentee is Ann, all remaining girls can be occupied if Beryl is supplied with limbless bodies; failing this, Cathie and Dot can be kept busy by a supply of legless bodies, or Dot alone by a supply of naked but otherwise complete dolls. In general, then, the choice between usable outside supplies is the greater the earlier along the line the absence occurred; when the last worker is absent only the finished product will do; when the first worker is absent all unfinished ones with the exception of the first will serve the purpose.

Synthesis of an Amino Acid by a Bacterium

One of the main constituents of living matter is protein. A protein molecule is a long chain of simpler organic molecules, called "amino acids." About twenty different amino acids are known to occur in proteins, and most organisms require all or most of them for life.

Higher organisms like man obtain their amino acids through the protein in their diet. Many microorganisms, for instance most bacteria, can make or "synthesize" at least some of the required amino acids from simple inorganic ingredients of their food. Where genetical methods could be employed in the study of amino acid synthesis that even bacteria can be studied genetically—it was found that amino acid synthesis proceeds in steps, each step being controlled by a different gene.

A, B, C, D = normal genes controlling successive steps in the synthesis of *AA* = an amino acid P_1, P_2, P_3 = intermediate products; precursors of *AA* = block in the reaction series where one of the genes is not functioning These genes, then, act in series like the girls at the conveyer belt. Our analysis of the "synthesis" of a doll will help us understand what happens when one of these genes mutates to an inactive state. Synthesizes the first step; gene *B*, comparable to Beryl, synthesizes the second; gene *C*, comparable to Cathie, synthesizes the third, and *D*, comparable to Dot, the fourth. *AA*, comparable to the dressed doll, is the finished amino acid; P_1, P_2 and P_3 are three intermediate products, or "precursors," comparable to the three unfinished stages of the doll. Bacteria in which *D* is not functioning can survive only when they are supplied with the finished amino acid; for *D*, like Dot, is the only one that can carry out the final step. Bacteria with an inactive gene *C* can survive on either the amino acid or its precursor P_3. Bacteria with an inactive gene *B* can use not only the amino acid and P_3 but also P_2.

Finally, bacteria in which A is not functioning can use all three precursors as well as the amino acid. Often such mutant strains accumulate precursors which precede the break in the production line; thus a strain with an inactive gene B may accumulate P_1, or a strain with an inactive gene *D*, P_3.

The study of chemical reactions in the living organism is called "biochemistry." The study of genes controlling biochemical reactions is called "biochemical genetics." The biochemical genetics of microorganisms has cleared up the manner in which many important compounds are synthesized under the control of genes acting in series.

There is a remarkable similarity in the biochemical pathways by which cells of widely differing organisms synthesize essential compounds such as amino acids and vitamins, and the biochemical genetics of microorganisms has contributed at least as much to biochemical as to genetical knowledge.

A gene-controlled Biochemical Sequence in Man

Lastly we shall consider an example from human genetics where genes have been shown to work in series not on the synthesis but on the breakdown of an amino acid. In 1908 an English physician, Garrod,called Inborn Errors of Metabolism.

In it he dealt with a number of inherited disorders of

human metabolism, that is, of those biochemical processes by which the living organism synthesizes new compounds or breaks down existing ones. One of the conditions described by Garrod was alkaptonuria, an abnormality characterized by a blackening of the urine and a blackening and hardening of the cartilage (gristle).

The disease is inherited as a recessive autosomal condition and must therefore be due to the malfunctioning of a single gene. When the urine of patients was analyzed it was found to contain a substance called alkapton, which is absent from the urine of normal persons. Now alkapton is one of the breakdown products of protein, or rather of one particular amino acid, called "phenylalanine."

When patients were given extra quantities of this amino acid in their diet, they excreted correspondingly more alkapton in their urine, while normal persons did not do so. In metabolism, phenylalanine is in the last resort broken down to carbon dioxide (CO_2) and water (H_2O). It appears that alkapton is an intermediate product in this chain of breakdown reactions, and that the gene for alkaptonuria interrupts the chain just after this stage has been reached.

Subsequently, a second gene—likewise autosomal and recessive —was found to intervene in the same chain of metabolic reactions, but at an earlier step. In this case, the mutation that led to the discovery of the gene has very serious effects; for homozygous individuals suffer from an extreme type of mental deficiency. The urine of these individuals contains large amounts of a substance called "phenylpyruvic acid," and the disease is called "phenylpyruvic idiocy" or "phenylketonuria".

In the metabolism of phenylalanine, phenylpyruvic acid is the first breakdown product. In patients it is, therefore, the first step in the metabolic reaction chain that is blocked by the mutated gene; in normal individuals, it is the first step that is mediated by the normal allele.

DNA

DNA, Like Protein, forms giant molecules which consist

of smaller and simpler components. The components of protein are called "amino acids", the components of nucleic acid are called "nucleotides." Each nucleotide, in turn, consists of three parts: a sugar molecule, a phosphate group, and a so called purine or pyrimidine base. Purines and pyrimidines are organic molecules of a kind found in many natural products; caffeine, for instance, is a purine.

All nucleotides of DNA contain the same sugar molecule, called "desoxyribose"; the nucleotides of RNA contain, instead, a sugar called "ribose," which differs from desoxyribose by having one additional oxygen atom. The bases in the different nucleotides of DNA are not all the same; there are four bases, two purines and two pyrimidines, and each nucleotide carries one of them. Rarely, a fifth base may occur.

To give the formulae of the purines and pyrimidines of DNA would take us too far into chemistry; but it will simplify the presentation of what follows if we know their chemical names. The two purines are called "adenine" (*A*) and "guanine" (*G*): the two pyrimidines are called "thymine" (*T*) and "cytosine" (*C*).

Originally, it was thought that a giant molecule of DNA always contained equal numbers of *A, G, T* and *C*; that, in fact, each such giant molecule could be subdivided into groups of four nucleotides, carrying between them just one *A*, one *G* and one *T* and one *C*. If this were correct, the potentialities of DNA as carrier of a genetical code would be rather limited; for—as can easily be veri fied—each group of four nucleotides could at best yield only $2 \times 3 \times 4 = 24$ different "letters," and many such groups of four would be required for the spelling of even a simple "word," although thousands of "words," that is, gene-controlled processes, have to be specified for the development of even the simplest organism. Until recently, these considerations inclined most geneticists to regard the protein component of the chromosome as the essential carrier of genetic information.

This concept of the structure of DNA has now been abandoned. The proportion between the four bases in a molecule of DNA differs between species, and it is not always

1: 1: 1: 1. There is no longer any reason to assume that the bases in a DNA molecule are arranged in groups of four; on the contrary, it is probable that they can be arranged in every possible sequence, such as *AATGCGA* or *TGTGCCA* and so on.

This at once does away with the limitations of coding. It is evident that even very short sequences of, say, three neighboring nucleotides can "spell" large varieties of different "letters" if all possible combinations and permutations among the four bases are allowed. Three nucleotides may, for instance, spell *AAA* or *CCC* or *AGA* or *TCA* and so on. For bacteria and viruses, where DNA has been shown to carry the genetic information, it is reasonable to assume that the code is spelled by the sequence of nucleotides, and it is at least probable that this is true also for higher organisms.

There remains, however, one peculiar numerical relationship among the four bases of DNA : the number of purines is always equal to the number of pyrimidines ($A+G = T+C$). Closer analysis shows that this matching between purines and pyrimidines goes even further, and that the number of adenines equals that of thymines ($A = T$), while the number of guanines equals that of cytosines ($G = C$). Thus, if in the DNA of a given species there is, for example, an excess of adenine over guanine, there is a corresponding excess of thymine over cytosine.

This chemical property of DNA remained unexplained until in 1953 two young scientists at Cambridge, the Englishman Crick and the American Watson, put forward a new idea for the structure of DNA. Their model not only explained the curious quantitative relationship among the bases, it also gave a simple interpretation of one of the most puz zling biological properties of the genetic material, its ability for accurate replication.

Before we consider the Watson-Crick model of DNA, let us look at the difficulties that confronted geneticists when they tried to understand gene replication. It is difficult to think of similar processes in nature which might provide a key to the mystery. When a crystal of, say, ordinary household salt is kept in a strong solution of the same salt it "grows" by the addition of salt molecules, and the new molecules are added on in such a way that the characteristic shape of the crystal—in this particular case a cube—is maintained. The resulting large crystal can easily be split again into smaller cubes.

There is a certain resemblance between gene replication and this process of growth and subdivision according to a fixed geometrical pattern, but the differences are much more striking than the similarities. A crystal has a simple molecular structure; its atoms are arranged in regular patterns. Genes—whatever their particular chemical structure—have such highly specific functions that they must differ from each other in complex and subtle ways.

It is these complexities and subtleties that are replicated faithfully every time a gene multiplies, and this cannot be achieved by the simple physical forces that govern the growth of a crystal. A better comparison can be made with antibody formation. Antigens resemble genes in their high degree of specificity, and they transfer specificity to the antibodies that they provoke.

The hypothesis that the procedure by which a cell makes "copies" of its genes is similar to that by which an antibody copies the specificity of the antigen is an attractive idea, but it has its difficulties. The antibody is not a copy of the antigen; their relationship to each other is that of the mold, or "template," for the stamping of a coin.

In order to replicate the antigen molecule itself, the antibody would have to serve as template for making an antigen, which then indeed would be exactly like the original one. Thus, theories of gene replication that are based on a comparison with antibody formation have to assume that the gene is the template for some intermediate product, which in turn is the template for the new gene.

The outstanding feature of the Watson Crick model is that it can explain gene replication without recourse to a hypothetical intermediate carrier of specificity.

Watson and Crick based their model on the X-ray diffraction picture of DNA. When X rays pass through a crystal they are scattered by its atoms, and from the pattern that the scattered beams produce on the X-ray plate an experienced scientist can infer the relative positions of the atoms to each other.

This technique has been taken over for substances that are not true crystals but whose molecules, like those of protein or nucleic acid, consist of regularly arranged and spaced subunits. X-ray diffraction work on nucleic acid has been carried out for many years.

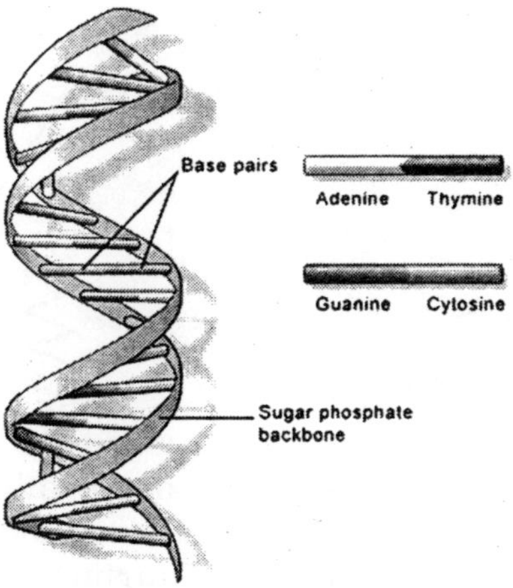

Fig. The Strucure of DNA

The DNA molecule, only part of which is shown, consists of two threads that are wound spirally round each other. Since the diameter of the windings remains the same throughout the molecule, it is more correct to speak of helical winding and to call the whole structure a double helix.

In reality, of course, the strands are not solid fibers but consist of individual atoms, held together by strong chemical bonds. Weaker bonds connect the two strands crosswise and hold them together. Chemically, the strands are formed by alternating units of phosphate and sugar; the bases are attached to the sugar, one base to each sugar unit, and they reach out into the hollow cylinder formed by the two winding strands.

Opposite bases are held together by weak chemical bonds.

If we imagine the strands unwound and the whole structure flattened out it would look somewhat like this: Now, of the two kinds of base, the purines are considerably larger than the pyrimidines. If a pyrimidine on one strand were paired with a pyrimidine on the other, the diameter of the whole structure in this region would be narrower than in a region where two purines were paired.

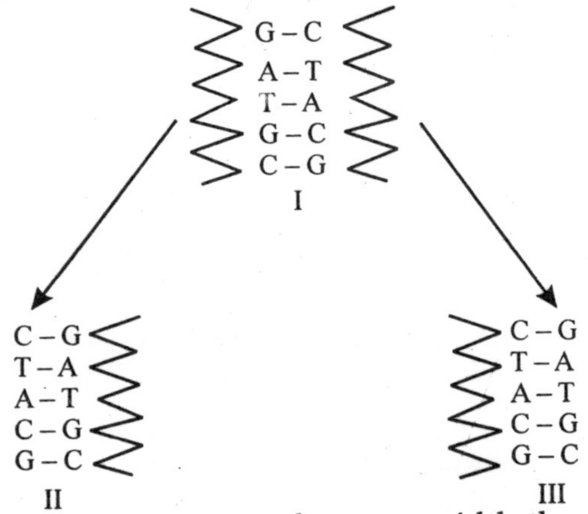

If one wants to preserve the same width throughout the whole DNA molecule, one has to arrange the bases in such a way that at every crosslink a purine is paired with a pyrimidine. Since, as the X-ray diffraction picture shows, the diameter of DNA does not change from region to region, this must be the way the bases are arranged in nature.

You will realise that this conclusion automatically results in equal proportions of purines and pyrimidines and explains one of the hitherto puzzling numerical relations among the bases. Moreover, when the chemical configurations of the individual bases are taken into account it turns out that, in order to fit into the available space, adenine has to pair with thymine, and guanine with cytosine.

This explains why there are always as many adenines as thymines and as many guanines as cytosines in a molecule of DNA. The base sequences in the two strands are thus complementary to each other: if we choose one of them arbitrarily, the other one is no longer open to choice. If, for instance, the sequence in a region of one strand is AGGTTCTGAC then the opposite sequence necessarily is TCCAAGACTG

It remains to discuss the greatest attraction of the Watson Crick model for the geneticist: its ability to provide a simple mechanism for gene replication. According to present day theory, DNA duplicates itself in the following way: First, the two strands unwind and separate; how this is achieved is still somewhat of a puzzle.

Once the strands are free, the base sequences, which according to our present view carry the genetical code, are exposed to the surrounding fluid, which contains all kinds of chemical building blocks, among them phosphate, sugar, and the four purine and pyrimidine bases required for synthesis of DNA.

From this store of building material each base of an original strand attaches to itself a new base, and these new bases then become linked with each other by a new sugar phosphate strand, which proceeds to wind round the old one and form a double helix with it. The most important point about the whole process is that the old bases are not free in the choice of the new ones.

Unless each base chooses its complementary one, that is, unless A chooses T, T chooses A, G chooses C, and C chooses G, there will be a bulge or a constriction in the new double helix wherever such a wrong choice has been made, and this poor fit will cause the unsuited base to be replaced by the fitting one.

This diagram shows the essential features of DNA replication, as envisaged by modern theory: each "daughter" chromosome is made up of one old and one new strand, and the specific base sequence in the two daughter molecules (II and III) is the same as in the original one.

Like every major discovery, the Watson Crick model has posed many more questions than it has answered, questions such as: Is the whole of the chromosomal DNA used for coding, or does a chromosome contain some "nonsense" sequences, which are used as full stops or commas or for some altogether different purpose? Is the genetical information restricted to DNA, even in higher organisms in which the chromosomal DNA is always tightly bound up with protein and a certain amount of RNA? How is the code deciphered and translated into specific biochemical processes? RNA obviously plays a major role in this; but how exactly does it function?

THE MONOPOLY OF THE GENES IN INHERITANCE

The transmission of a character depends on genes, although its manifestation may be influenced by environment. Since the genes are in the nucleus, this type of inheritance is called "nuclear." The overwhelming majority of well studied cases of inheritance are nuclear, but this does not mean that "extranuclear" or "cytoplasmic" inheritance is impossible.

In fact, we have already encountered one case of extranuclear inheritance. Chlorophyll formation, as we saw, is under the control of a large number of genes; but these genes require the co-operation of the chloroplast. Changes may occur in a chloroplast that make it unable to respond to the controlling genes, and these changes may be transmitted from the first abnormal chloroplast to all its descendants.

Many cases of chloroplast inheritance are known, and since chloroplasts—like genes—reproduce by replication and division, this type of extranuclear inheritance is fairly well understood.

Other cases of cytoplasmic inheritance exist and can be detected by special methods. Two rules are used for their detection, one negative, one positive. The negative rule states that hereditary traits which are transmitted through the cytoplasm must not show Mendelian segregation in crosses. The positive rule is based on the fact that, at least in all higher organisms, the female gamete contains a large amount of cytoplasm, while the male gamete contains very little.

One would therefore expect that cytoplasmically inherited characters would follow the maternal line. This is, indeed, true for chloroplast inheritance.

In a cross between plants with normal and abnormal chloroplasts, *all* the progeny have normal chloroplasts if those of the mother plant were normal and abnormal ones if those of the mother plant were abnormal. Occasionally, some chloroplasts may be transmitted through the pollen, but this does not alter the essential difference between the progeny of the two "reciprocal" crosses.

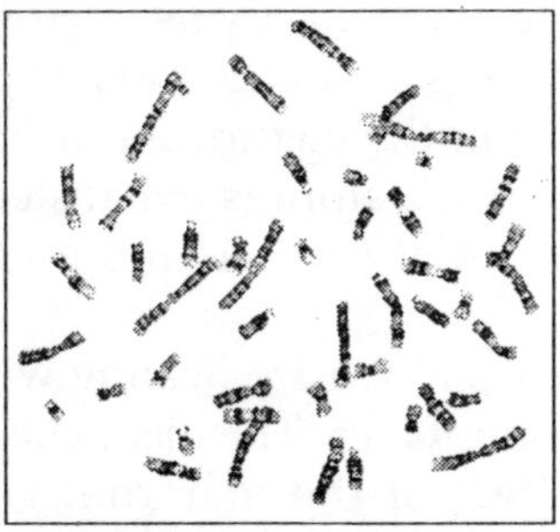

Fig. Genes of Inheritance

A comparison between the results of reciprocal crosses is the most useful tool in the search for cytoplasmic inheritance. But it is a tool that has to be used with caution, for differences between progenies of reciprocal matings may occur also when inheritance is purely nuclear.

The simplest example is sex-linked inheritance. A cross between a gold cock *ss* and a silver hen *S* produces silver cockerels *Ss* and gold pullets; the reciprocal cross between a silver cock *SS* and a gold hen *s* produces only silver offspring: the cockerels are *Ss*, the pullets are *S*. Such cases are easily distinguished from cytoplasmic inheritance.

Another cause for differences between reciprocal crosses is early gene action in the unfertilized ovum. In the flour moth Ephestia, the normal black eye colour of the caterpillars and moths develops through a series of biochemical steps, one of which is blocked when a gene *A* is replaced by its allelomorph *a*. *A* is dominant over *a*, and the eyes of *Aa* caterpillars and moths are black like those of homozygotes *AA*.

The eyes of *aa* animals are light because one of the necessary precursor substances is not produced in the absence of *A*. When a heterozygous black eyed male *Aa* is crossed with a light eyed female *aa*, the progeny segregates into 50 per cent *Aa* caterpillars with black eyes and 50 per cent *aa* caterpillars with light eyes, just as could have been predicted from Mendel's first law. The reciprocal cross gives a different result. Here all caterpillars have black eyes. That segregation into *Aa* and *aa* did, nevertheless, occur can be seen when the moths emerge from their cocoons; for half of them have black eyes and half have white eyes, exactly as in cross I.

Apparently the *A* gene in the mother produces sufficient precursor substance in the egg to last through the lifetime of the caterpillar; when the store is exhausted, no more can be formed in the absence of gene *A*, and no black eye pigment develops in the moth.

Again it is clear that we are dealing with nuclear and not with cytoplasmic inheritance. The distinction would have been less easy if the effect of the maternal gene had persisted throughout the life of the progeny. This, too, may happen and gives rise to cases that at first sight look like cytoplasmic inheritance.

Whether an individual will develop a left handed or a right handed coil depends on the arrangement of the first four cells in the young embryo, and this in turn depends on a pattern of asymmetry that is laid down in the unfertilized egg under the influence of the mother's genes.

One pair of genes controls this pattern, *D* the gene for right handed coiling being dominant over its recessive allele d for left handed coiling. A mother of genotype *DD* or *Dd* has only right handed offspring, independent of the genotype of the father.

Similarly, a mother of genotype *dd* has only left handed offspring. In general, the direction of coiling in any individual is determined by the genotype of its mother, not by its own genotype. This can easily be followed in the breeding scheme, in which use has been made of the fact that snails are hermaphroditic, that is, they have both male and female sex organs and are capable of either self fertilization or cross fertilization.

If the breeding tests had not been continued beyond the first generation, this would have seemed a clear case of cytoplasmic inheritance according to our rules; for there was no segregation, and in reciprocal crosses all offspring had the maternal phenotype.

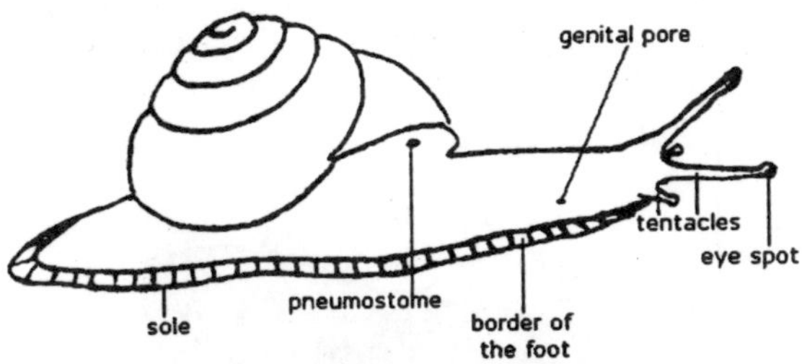

Fig. Pond Snail Limnaea

When all similar instances of "maternal inheritance" of nuclear effects are left aside, there remain a number of cases that must be attributed to cytoplasmic inheritance, or at least to inheritance by way of the cytoplasm. This is not a purely verbal distinction; it refers to the experience that several cases of what appeared to be cytoplasmic inheritance turned out to be transmission of a microorganism through the cytoplasm.

Perhaps the most spectacular case is that of the "killers" and "mate killers" in the tiny fresh water animal Paramecium. This organism consists of one cell and produces clones by continued division into two; in addition, mating occurs and gives Mendelian segregation of genes.

When different clones of this animal are kept together in a container, it may happen that all those of one clone are killed by a secretion from the other.

For the killers themselves, this secretion is harm less. Different killer strains have different means of killing their victims. Most of them do not kill their mates; but there an some strains that, instead of killing from a distance by secretion, kill their mates through close contact.

The killer character has a nuclear as well as a cytoplasmic basis. The situation is very similar for ordinary killers and mate killers; it will be described for the former. The nuclear basis is a pair of alleles *K* and *k*. Animals that are homozygous for *k* are sensitive to killing and cannot themselves become killers.

Animals that carry the *K* homozygously or heterozygously in normal cytoplasm are potential killers; they are actual killers when their cytoplasm contains particles called "kappa," which in turn produce the lethal poison. In *KK* or *Kk* animals kappa particles are transmitted from cell to cell in the cytoplasm; once they have been lost from a cell they do not again develop by themselves.

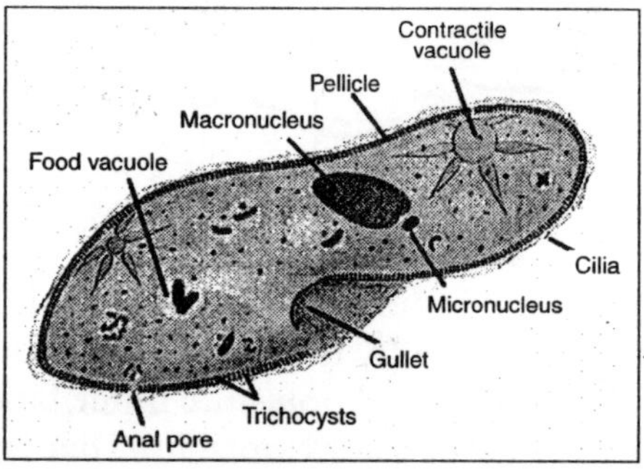

Fig. The Microscopic animal Paramecium

The large bean shaped body and the small round one adjoining it are nuclei. The star shaped structures are vacuoles which pump superfluous fluid out of the cell. The inheritance of kappa was at first considered a good example of cytoplasmic inheritance. Closer studies, however, showed that the kappa particles resemble bacteria, and their transmission from cell to cell in the cytoplasm is therefore more aptly compared with the transmission of a parasitic microorganism.

The fact that kappa particles can be maintained only in animals with the gene K is no argument against this interpretation. It is a very common observation that resistance or sensitivity to a parasite may depend on the genotype of the infected host. In man there is evidence for genetically determined sensitivity to the polio virus, and bacterial strains may become resistant to a bacteriophage through mutation of one of their genes.

In Drosophila, certain animals are hypersensitive to the effects of carbon dioxide and are killed by doses that have only slight and transitory effects on normal flies. This property is inherited cytoplasmically; but, again, the responsible particles appears to be a virus rather than a normal component of cytoplasm.

A decision between these alternatives is not easy, and geneticists are divided in their interpretation of this case and similar ones. But even if all of them are attributed to true cytoplasmic inheritance, their number among animals is exceedingly small and is not likely to be increased considerably by future discoveries.

It is somewhat larger in plants, where reciprocal crosses often differ and the differences do not always regard the chloroplasts. One character that in many plants is inherited through the cytoplasm has a certain economic importance. This is male sterility, the inability to form functional pollen. Where, as in the breeding of hybrid corn, selffertilization has to be prevented, the introduction of the cytoplasmic factor for male sterility into the line used as female parent can save much work.

In summary, it seems safe to conclude that in the control of inheritance the monopoly of the genes, although not absolute, is very strong. Whether cytoplasmic inheritance plays a role in determining differences between widely differing organisms, belonging to different species or genera, is a much debated question.

Unfortunately, it is not amenable to experimental solution, because crosses between species or genera are usually sterile or, if fertile, have sterile progeny (like the mule). We shall not consider cytoplasmic inheritance when, we shall throw a brief glance at the forces that have brought about evolution.

MUTATION

An observed mutation reflects a change in a gene. Presumably the mutated gene differs from the original one in chemical composition or structure. The new gene, like the old one, must be capable of accurate self replication, for a mutated gene replicates in the mutated form. There is no recovery from mutation except in the very rare cases when a subsequent mutation restores the original gene. If our present view of the nature of the gene is correct, a mutation consists in a change in the number, types, or arrangement of nucleotides that "spell" a gene. On the Watson Crick model of DNA replication, such changes would be perpetuated. Mutation is not restricted to the germ cells; however, in sexually reproducing species only mutations that occur in germ cells are transmitted to the progeny.

In asexually propagated species, mutations that occur in ordinary body cells may become established as mutant strains: many varieties of fruit trees have originated as "bud sports." In a diploid species, recessive mutations can produce bud sports or mosaics only when the individual is already heterozygous for the gene in question.

A snapdragon flower which is half purple and half lavender; it grew on a branch which, in addition to two or more such mosaic flowers, also bore several that were either wholly purple or wholly lavender. The plant itself was heterozygous for the recessive lavender gene, and in one of the cells from which the mosaic branch developed the normal allele had mutated to lavender. Dominant mutations may give rise to immediately observable bud sports or mosaics.

Mutation takes place regularly in all species of organisms. Each individual gene mutates only very rarely, perhaps once in 100,000 or once in a million cells; but as the number of genes in most organisms is very high the over all mutation frequency per generation may be considerable. In man some mutations, for instance, that to hemophilia, occur at the relatively high frequency of about one in 50,000; this means that about one out of 50,000 gametes carries a newly arisen gene for hemophilia.

For various reasons, it is likely that the majority of human genes mutate less frequently, perhaps once in 100,000 gametes or even more rarely. If we make the conservative assumption that the 23 chromosomes in a human gamete between them carry 10,000 genes, then it would follow that one gamete out of ten carries a newly mutated gene. This is not more than a rough guess, but it is not likely to be too high.

Only a small fraction of new mutations is immediately detectable as mutant offspring. A recessive mutation in the gamete of a diploid species will yield phenotypically normal offspring, unless the other gamete happens to carry the some recessive gene as mutation in the population. Apart from such cases, a newly arisen recessive gene may be carried in heterozygotes for many generations, until the mating of two such heterozygotes produces a homozygous offspring. An exception is formed by sexlinked recessive mutations; these will show up in the first XY individual with the mutated gene.

Even dominant or partially dominant mutations do not always give rise to detectably mutant progeny. Many of them produce effects that are too small to be observed or that might equally well be attributed to environmental influences. This applies particularly to mutations affecting quantitative characters, for instance, egg production in poultry, grain yield in corn, general vigour, or intelligence.

Thus the realization that about one out of ten human gametes carries a new mutation does not mean that about one out of ten children will be in some way abnormal or unusual. It does, however, mean that in man, as in every other species, new mutations occur inexorably even without artificial causes such as exposure to X rays.

These new mutations must not be thought of as new in the sense that they have not occurred before. A species has a limited number of genes, and each gene can mutate only to a limited number of alleles. Each mutation of a given gene to a given allele recurs at a certain low frequency, and in a species with a long evolutionary history every possible mutation is likely to have occurred repeatedly. In microorganisms, in which millions of cells can be screened for mutations, it is easy to see that the same mutations turn up again and again.

This is an important fact. It explains why the majority of mutations are harmful. In a species that, through long periods of evolution, has become adapted to its mode of life all useful alleles that can arise by mutation will already have been incorporated into the genotype, so that all or most members of the species carry them in homozygous condition.

Once this stage has been reached, only harmful or at least less valuable alleles can still occur as new mutations. Most desert animals have a yellow gray coat colour, which forms an effective camouflage. Evolution in the desert has favored animals with the right type of coat colour genes, and alleles arising from new mutations can only be to lighter or darker shades, away from the best possible one.

This is a simple example, but the same principle applies to more complicated cases in which many genes interact to produce the best adapted phenotype.

Indeed, the more complex the genetic control of a character the more readily is it upset by mutation in one of its controlling genes. A primitive piece of machinery may continue to work when one of its pieces has been replaced by a slightly different one; a complicated apparatus will be wrecked by the slightest alteration in the dimensions or position of one of its cogs or levers. Yet even the most complicated man made machine is simple compared with the most primitive living organism.

The fact that the majority of mutations is harmful has recently developed into a matter of immediate and grave concern. As long as mutations occur only from natural sources, there is nothing we can do about controlling their frequency. Modern man, however, is making use of agents that artificially increase the frequency of mutations above its natural level.

Foremost among these agents are high energy radiations from X-ray machines, from fallout produced by nuclear tests, and from a number of other sources. The benefits to be derived from these devices have to be balanced against the harm we inflict on unborn generations by loading them with an increased burden of mutated genes. So much has been written about this urgent and difficult problem that it would be redundant to discuss it here.

Only one point may be mentioned. The question is often raised whether new mutations may not sometimes be beneficial. This is, indeed, true. Very occasionally beneficial mutations occur. Plant breeders have successfully used radiation or chemical treatment for obtaining useful mutations in cereals, ornamental plants, fruit trees, and other plants. Irradiation of molds like Penicillium has produced strains with increased yield of antibiotic.

But each such useful mutation has been obtained at the

cost of about a thousand harmful or useless ones, which had to be discarded. Even in animal breeding such a procedure would be impossibly wasteful. For mankind any increase in mutation frequency remains highly undesirable in spite of the possibility that among a thousand harmful mutations there may be one that is beneficial.

While the genetical dangers of radiation are among the most widely discussed topics of our time, much less attention is paid to the possibility that mutations may be produced by some of the many chemicals that civilized man uses in his food, his drugs, his cosmetics, his industrial processes. There is a good reason for the lack of scientific pronouncement on this question: our profound ignorance of the relevant facts. It is true that the past twenty years have seen the discovery of an ever increasing number of chemicals that produce mutations in Drosophila, in plants, and in microorganisms.

But while it may be confidently assumed that human genes, like those of every other tested species including mice, will mutate on exposure to ionizing radiation, the same cannot be taken for granted in regard to chemical "mutagens." X rays penetrate easily through living matter of every description. Chemicals may be retained by the outer layers of the body before they reach the germ cells; they may be excreted before they have had a chance to produce mutations; they may be trapped by cell constituents outside the nucleus; they may be metabolized into compounds with different chemical properties.

All these possibilities will differ between species, and a chemical that produces many mutations in one organism may be quite ineffective in another. Thus caffeine has produced mutations in certain bacteria and fungi. In Drosophila it has had a similar, but much weaker, effect.

Its effect on the germ cells of mice, rats or guinea pigs remains to be tested by experiment. Whether it is a mutagen for people who imbibe it in coffee or tea cannot be decided experimentally. Data on mice and other laboratory mammals can provide no more than presumptive evidence, which different geneticists will interpret differently according to their scientific caution and their degree of addiction to coffee or tea. Yet, to obtain such data remains an urgent object for genetical research in the near future.

Mutation research thus forms an important branch of applied genetics, both positively, as a tool for the production of improved strains of plants and microorganisms, and negatively, as a means for detecting possible harmful genetical effects of processes and chemicals used in human societies. For the research worker mutation work is one of the most valuable tools in the study of the genetic material. When in 1927, at the Fifth International Congress of Genetics in Berlin, H. J. Muller from the United States first reported the production of mutations in Drosophila by X rays he opened in the history of genetics.

With the aid of X-ray machines, every laboratory worker could now produce unprecedented numbers of new mutations in whatever organism he was studying. These mutations, in turn, provided material for all types of genetical investigation, which without this might have had to wait for decades until the right kind of mutation turned up spontaneously. Nowadays the geneticist who wants some particular mutation for his experiments turns quite naturally to the X-ray machine to provide it.

In research on microorganisms, ultraviolet radiation serves the same purpose more effectively. Most of the mutations for the study of gene action in microorganisms or of genetic recombination in bacteria were produced by exposure of cells to ultraviolet light. In higher organisms, too, ultraviolet light can produce mutations when it reaches the nucleus of the germ cells; but shielding through the outer layers of tissue is so great that special tricks have to be used to ensure penetration. Although excessive exposure of the human body to the ultraviolet light of the sun may cause nasty burns, we need not worry about its effect on the progeny.

The first chemicals with mutagenic action were discovered at the beginning of the Second World War. At present we know scores of such substances, belonging to very different chemical groups. Some, like mustard gas and certain related substances, have produced mutations in every kind of tested organism. Others act only on certain types of organism or cell. Formaldehyde, for instance, produces mutations in Drosophila males but not in Drosophila females. In recent years much effort has been directed toward finding chemicals that are even more selective and act only on certain genes. Obviously the detection of such substances would be of tremendous importance for the theory and practice of genetics.

For the research worker they would provide a tool for studying the chemical nature of individual genes; for the applied geneticist they would provide means for producing at will mutations of a desirable type. We are still far from having reached this goal, but there have been encouraging signs that it may not be unattainable. It is already clear that chemicals may differ from radiation and from each other in the relative frequencies with which they produce different mutations.

We still know little about the manner in which radiation or chemical mutagens produce their effects. X rays and similar types of radiation release large amounts of energy on their passage through living matter; when this happens in or near a chromosome it may result in chemical change of a gene. The much smaller amounts of energy released by ultraviolet radiation may act in a similar way; but possibly much of the mutagenic action of ultraviolet light is mediated by chemical effects of the radiation on the fluid surrounding the genes. To some extent this appears to be true even for high energy radiations like X rays. Mutagenic chemicals may react with the genes directly, or indirectly through reactions with the cytoplasm. It is also possible—and indeed may be a frequent cause of mutation—that a gene "makes a mistake" when it replicates. One can imagine many ways by which a mutagen may provoke such mistakes, Especially mutagens that are chemically related to the purines and pyrimidines in DNA are liable to act in this way.

Our ideas about what causes mutations to occur in nature are even more speculative. Certainly radiation from cosmic rays and from radioactive substances in the soil and the organisms themselves must contribute their share of mutations; but the amount of radiation from these sources is too low to account for more than a small part of all naturally occurring mutations. Mutagenic chemicals may be responsible for much or all of the remainder. Several known mutagens are natural products; quite recently, a very potent mutagen—heliotrin—has been found in the common ragwort. In bacteria, at least, a proportion of naturally occurring mutations seems to be due to mutagenic purines that are produced in the cells themselves.

This can be concluded from experiments in which purines such as caffeine or related substances were used for producing mutations.

The mutagenic effect of the purines could be completely prevented when the bacteria were given certain other substances, which in some way acted as "antimutagens." When untreated bacteria were grown in solutions of these antimutagens they developed many fewer mutations than bacteria grown in ordinary nutrient solution. As would be expected, the antimutagens did not interfere with the production of mutations by radiation.

After these speculations we reach firm ground again when we consider a different kind of genetical change which occurs infrequently in nature and can be induced by radiation and many chemicals: chromosome breakage. In organisms with sufficiently large chromosomes it is easy to see breakage by X rays, mustard gas, or other mutagenic treatment; in other organisms breakage may be detected by genetical methods, that is, by examination of the progeny.

The genetical consequences of chromosome breakage depend on the fate of the broken chromosomes. In some way which is not yet fully understood broken chromosome ends can stick together to form again permanent structures; but, as chromosomes in a living cell float freely in the nuclear sap, the new chromosomes are not always the same as the original ones.

Chromosomes may exchange pieces, fragments may be lost, and other kinds of "structural change" may take place.

In the present context the most important change is a "deficiency," that is, loss of a fragment either from the end of a chromosome or, through rejoining between the wrong pieces, from one of its inner regions. A fragment that fails to rejoin usually is lost, and with it the genes carried on it. In a haploid organism like Chlamydomonas or many strains of yeast and other fungi all but the smallest deficiencies result in death, the vast majority of genes are necessary for survival. Either deficiency kills homozygotes, that is, individuals lacking the same gene or genes in both partner chromosome; but, while the deficiency has no noticeable.

Two breaks have occurred in the same chromosome. The broken ends have stuck together, the middle piece has been left out and is likely to be lost at one of the next cell divisions effect on the heterozygote, the deficiency makes even the heterozygote strikingly abnormal. Thus the first deficiency behaves like a recessive gene, the second like a dominant one.

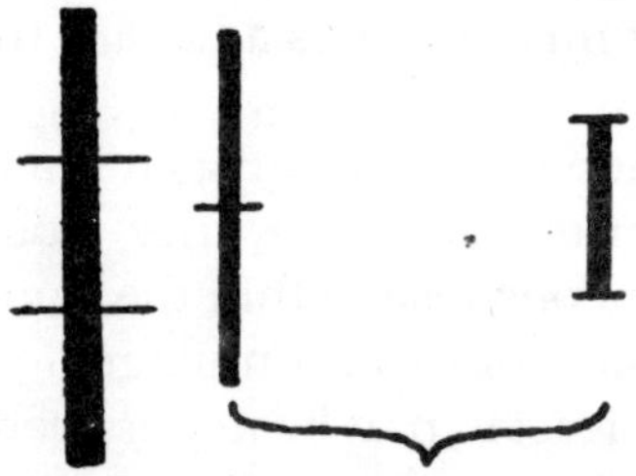

Fig. Origin of a Chromosome Deficiency

Cells with broken chromosomes, whether or not these rejoin into rearrangements, survive well as long as they do not divide, but for various reasons they usually die at the next division or soon after. This explains why dividing cells are much more sensitive to radiation than nondividing ones. The selective killing of dividing cells by X rays is utilized in the treatment of cancer. Tumors characteristically contain a much higher proportion of dividing cells than the surrounding normal tissue and are therefore selectively destroyed by radiation.

Treatment with mutagenic chemicals is of more recent origin. These chemicals, like X rays, destroy tumor cells by chromosome breakage; they may give good results especially in cases where X rays cannot be used. The first chemical of this kind was nitrogen mustard. Its value for the treatment of cancer was discovered in the United States at the beginning of the Second World War. Simultaneously and independently the first chemically induced mutations in Drosophila were found in Scotland in the progeny of flies that had been exposed to mustard gas or nitrogen mustard.

Also at the same time, and without knowledge of the American and British work, German scientists found that an entirely different substance—urethane—causes chromosome breakage and rearrangement in plants, and British clinicians tested urethane successfully on patients with leukemia.

This story of apparent coincidences is an illustration of the truth, well known to students of the history of science, that certain times are "ripe" for the discovery of certain scientific facts. Mendel was ahead of his time, and his great discovery remained practically unknown until the time had become ripe for it. This kind of misfortune is a hazard that is peculiar to a genius.

For the common run of scientists it is a humbling thought that in their absence their discoveries would probably have been made by someone else within the same decade.

While the ability of certain mutagenic agents to destroy tumor cells may at least in part be understood as a consequence of their ability to break chromosomes, another connection between mutagenesis and cancer is less easily interpreted. It is well known that X rays not only can *cure* cancer but that continued exposure to low doses of X rays may *produce* cancer, as the early X-ray workers learned to their great detriment. The same parallelism has meanwhile been found for a number of chemical mutagens, including nitrogen mustard and urethane.

Index

Standard Mendelian crosses were made between the two species and the occurence of diapause in the hybrids was